Lecture Notes in Mathematics

Editors:
A. Dold, Heidelberg
B. Eckmann, Zürich
F. Takens, Groningen

Subseries:
Mathematisches Institut der Universität
und Max-Planck-Insitut für Mathematik,
Bonn - vol. 19

Advisor:
F. Hirzebruch

Lothar Göttsche

Hilbert Schemes of Zero-Dimensional Subschemes of Smooth Varieties

Springer-Verlag

Berlin Heidelberg New York
London Paris Tokyo
Hong Kong Barcelona
Budapest

Author

Lothar Göttsche
Max-Planck-Institut für Mathematik
Gottfried-Claren-Str. 26
53225 Bonn, Germany

Mathematics Subject Classification (1991): 14C05, 14N10, 14D22

ISBN 3-540-57814-5 Springer-Verlag Berlin Heidelberg New York
ISBN 0-387-57814-5 Springer-Verlag New York Berlin Heidelberg

© Springer-Verlag Berlin Heidelberg 1994
Printed in Germany

SPIN: 10078819   46/3140-543210 - Printed on acid-free paper

Introduction

Let X be a smooth projective variety over an algebraically closed field k. The easiest examples of zero-dimensional subschemes of X are the sets of n distinct points on X. These have of course length n, where the length of a zero-dimensional subscheme Z is $dim_k H^0(Z, \mathcal{O}_Z)$. On the other hand these points can also partially coincide and then the scheme structure becomes important. For instance subschemes of length 2 are either two distinct points or can be viewed as pairs (p, t), where p is a point of X and t is a tangent direction to X at p.

The main theme of this book is the study of the Hilbert scheme $X^{[n]} :=$ $\text{Hilb}^n(X)$ of subschemes of length n of X; this is a projective scheme parametrizing zero-dimensional subschemes of length n on X. For $n = 1, 2$ the Hilbert scheme $X^{[n]}$ is easy to describe; $X^{[1]}$ is just X itself and $X^{[2]}$ can be obtained by blowing up $X \times X$ along the diagonal and taking the quotient by the obvious involution, induced by exchanging factors in $X \times X$.

We will often be interested in the case where $X^{[n]}$ is smooth; this happens precisely if $n \leq 3$ or $\dim X \leq 2$. If X is a curve, $X^{[n]}$ coincides with the nth symmetric power of X, $X^{(n)}$; more generally, the natural set-theoretic map $X^{[n]} \rightarrow X^{(n)}$ associating to each subscheme its support (with multiplicities) gives a natural desingularization of $X^{(n)}$ whenever $X^{[n]}$ is smooth.

The case $\dim X = 2$ is particularly important as this desingularization turns out to be crepant; that is, the canonical bundle on $X^{[n]}$ is the pullback of the dualizing sheaf or $X^{(n)}$ (in particular $X^{(n)}$ has Gorenstein singularities). In this case, $X^{[n]}$ is an interesting $2n$-dimensional smooth variety in its own right. For instance, Beauville [Beauville (1),(2),(3)] used the Hilbert scheme of a K3-surface to construct examples of higher-dimensional symplectic manifolds.

One of the main aims of the book is to understand the cohomology and Chow rings of Hilbert schemes of zero-dimensional subschemes. In chapter 2 we compute Betti numbers of Hilbert schemes and related varieties in a rather general context using the Weil conjectures; in chapter 3 and 4 the attention is focussed on easier and more special cases, in which one can also understand the ring structure of Chow and cohomology rings and give some enumerative applications.

In chapter 1 we recall some fundamental facts, that will be used in the rest of the book. First in section 1.1, we give the definition and the most important properties of $X^{[n]}$; then in section 1.2 we explain the Weil conjectures in the form in which we are later going to use them in order to compute Betti numbers of Hilbert schemes, and finally in section 1.3 we introduce the punctual Hilbert scheme, which parametrizes subschemes concentrated in a point of a smooth variety. We hope that the non-expert reader will find in particular sections 1.1 and 1.2 useful as a quick reference.

In chapter 2 we compute the Betti numbers of $S^{[n]}$ for S a surface, and of

KA_{n-1} for A an abelian surface, using the Weil conjectures. Here KA_{n-1} is a symplectic manifold, defined as the kernel of the map $A^{[n]} \to A$ given by composing the natural map $A^{[n]} \to A^{(n)}$ with the sum $A^{(n)} \to A$; it was introduced by Beauville [Beauville (1),(2),(3)].

We obtain quite simple power series expressions for the Betti numbers of all the $S^{[n]}$ in terms of the Betti numbers of S. Similar results hold for the KA_{n-1}. The formulas specialize to particularly simple expressions for the Euler numbers of $S^{[n]}$ and KA_{n-1}. It is noteworthy that the Euler numbers can also be identified as the coefficients in the q-development of certain modular functions and coincide with the predictions of the orbifold Euler number formula about the Euler numbers of crepant resolutions of orbifolds conjectured by the physicists. The formulas for the Betti numbers of the $S^{[n]}$ and KA_{n-1} lead to the conjecture of similar formulas for the Hodge numbers. These have in the meantime been proven in a joint work with Wolfgang Soergel [Göttsche-Soergel (1)]. One sees that also the signatures of $S^{[n]}$ and KA_{n-1} can be expressed in terms of the q-development of modular functions. The formulas for the Hodge numbers of $S^{[n]}$ have also recently been obtained independently by Cheah [Cheah (1)] using a different technique.

Computing the Betti numbers of $X^{[n]}$ can be viewed as a first step towards understanding the cohomology ring. A detailed knowledge of this ring or of the Chow ring of $X^{[n]}$ would be very useful, for instance in classical problems in enumerative geometry or in computing Donaldson polynomials for the surface X.

In section 2.5 various triangle varieties are introduced; by triangle variety we mean a variety parametrizing length 3 subschemes together with some additional structure. We then compute the Betti numbers of $X^{[3]}$ and of these triangle varieties for X smooth of arbitrary dimension, again by using the Weil conjectures.

The Weil conjectures are a powerful tool whose use is not as widely spread as it could be; we hope that the applications given in chapter 2 will convince the reader that they are not only important theoretically, but also quite useful in many concrete cases.

Chapters 3 and 4 are more classical in nature and approach then chapter 2. Chapter 3 uses Hilbert schemes of zero-dimensional subschemes to construct and study varieties of higher order data of subvarieties of smooth varieties. Varieties of higher order data are needed to give precise solutions to classical problems in enumerative algebraic geometry concerning contacts of families of subvarieties of projective space. The case that the subvarieties are curves has already been studied for a while in the literature [Roberts-Speiser (1),(2),(3)], [Collino (1)], [Colley-Kennedy (1)]. We will deal with subvarieties of arbitrary dimension and construct varieties of second and third order data. As a first application we compute formulas for the numbers of higher order contacts of a smooth projective variety with linear subvarieties in the ambient projective space. For a different and more general construction, which is however also more difficult to treat, as well as for examples of the type of problem that can be dealt with, we also refer the reader to [Arrondo-Sols-Speiser (1)].

The last chapter is the most elementary and classical of the book. We describe the Chow ring of the relative Hilbert scheme of three points of a $\mathbf{P}^2$ bundle. The main example one has in mind is the tautological $\mathbf{P}^2$-bundle over the Grassmannian of two-planes in $\mathbf{P}^n$. In this case it turns out hat our variety is a blow up of $(\mathbf{P}^n)^{[3]}$. This fact has been used in [Rosselló (2)] to determine the Chow ring of $(\mathbf{P}^3)^{[3]}$. The techniques we use are mostly elementary, for instance a study of the relative Hilbert scheme of finite length subschemes in a $\mathbf{P}^1$-bundle; I do however hope that the reader will find them useful in applications.

For a more detailed description of their contents the reader can consult the introductions of the chapters.

The various chapters are reasonably independent from each other; chapters 2, 3 and 4 are independent of each other, chapter 2 uses all of chapter 1, chapter 3 uses only the sections 1.1 and 1.3 of chapter 1 and chapter 4 uses only section 1.1.

To read this book the reader only needs to know the basics of algebraic geometry. For instance the knowledge of [Hartshorne (1)], is certainly enough, but also that of [Eisenbud-Harris (1)] suffices for reading most parts of the book. At some points a certain familiarity with the functor of points (like in the last chapter of [Eisenbud-Harris (1)]) will be useful. Of course we expect the reader to accept some results without proof, like the existence of the Hilbert scheme and obviously the Weil conjectures.

The book should therefore be of interest not only to experts but also to graduate students and researchers in algebraic geometry not familiar with Hilbert schemes of points.

Acknowledgements

I want to thank Professor Andrew Sommese, who has made me interested in Hilbert schemes of points. While I was still studying for my Diplom he proposed the problem on Betti numbers of Hilbert schemes of points on a surface, with which my work in this field has begun. He also suggested that I might try to use the Weil conjectures. After my Diplom I studied a year with him at Notre Dame University and had many interesting conversations. During most of the time in which I worked on the results of this book I was at the Max-Planck-Institut für Mathematik in Bonn. I am very grateful to Professor Hirzebruch for his interest and helpful remarks. For instance he has made me interested in the orbifold Euler number formula. Of course I am also very grateful for having had the possibility of working in the inspiring atmosphere of the Max-Planck-Institut.

I also want to thank Professor Iarrobino, who made me interested in the Hilbert function stratification of $\mathrm{Hilb}^n(k[[x,y]])$. Finally I am very thankful to Professor Ellingsrud, with whom I had several very inspiring conversations.

Contents

Contents

1. Fundamental facts

In this work we want to study the Hilbert scheme $X^{[n]}$ of subschemes of length n on a smooth variety. For this we have to review some concepts and results. In [Grothendieck (1)] the Hilbert scheme was defined and its existence proven. We repeat the definition in paragraph 1.1 and list some results about $X^{[n]}$. $X^{[n]}$ is related to the symmetric power $X^{(n)}$ via the Hilbert-Chow morphism $\omega_n : X^{[n]}_{red} \longrightarrow X^{(n)}$. We will use it to define a stratification of $X^{[n]}$. In chapter 2 we want to compute the Betti numbers of Hilbert schemes and varieties that can be constructed from them by counting their points over finite fields and applying the Weil conjectures. Therefore we give a review of the Weil conjectures in 1.2. Then we count the points of the symmetric powers $X^{(n)}$ of a variety X, because we will use this result in chapter 2. In 1.3 we study the punctual Hilbert scheme $\mathrm{Hilb}^n(k[[x_1, \ldots, x_d]])$, parametrizing subschemes of length n of a smooth d-dimensional variety concentrated in a fixed point. In particular we give the stratification of Iarrobino by the Hilbert function of ideals.

1.1. The Hilbert scheme

Let T be a locally noetherian scheme, X a quasiprojective scheme over T and $\mathcal{L}$ a very ample invertible sheaf on X over T.

Definition 1.1.1. [Grothendieck (1)] Let $\mathcal{H}ilb(X/T)$ be the contravariant functor from the category $\underline{\mathrm{Schln}}_T$ of locally noetherian T-schemes to the category $\underline{\mathrm{Ens}}$ of sets, which for locally noetherian T-schemes U, V and a morphism $\phi : V \longrightarrow U$ is given by

$$\mathcal{H}ilb(X/T)(U) = \left\{ Z \subset X \times_T U \text{ closed subscheme, flat over } U \right\}$$
$$\mathcal{H}ilb(X/T)(\phi) : \mathcal{H}ilb(X/T)(U) \longrightarrow \mathcal{H}ilb(X/T)(V); Z \longmapsto Z \times_U V.$$

Let U be a locally noetherian T-scheme, $Z \subset X \times_T U$ a subscheme, flat over U. Let $p : Z \longrightarrow X$, $q : Z \longrightarrow U$ be the projections and $u \in U$. We put $Z_u = q^{-1}(u)$. The Hilbert polynomial of Z in u is

$$P_u(Z)(m) := \chi(\mathcal{O}_{Z_u}(m)) = \chi(\mathcal{O}_{Z_u} \otimes_{\mathcal{O}_z} p^*(\mathcal{L}^m)).$$

$P_u(Z)(m)$ is a polynomial in m and independent of $u \in U$, if U is connected. For every polynomial $P \in \mathbf{Q}[x]$ let $\mathcal{H}ilb^P(X/T)$ be the subfunctor of $\mathcal{H}ilb(X/T)$ defined by

$$\mathcal{H}ilb^P(X/T)(U) = \left\{ \begin{array}{c|c} Z \subset X \times_T U & Z \text{ is flat over } U \text{ and} \\ \text{closed subscheme} & P_u(Z) = P \text{ for all } u \in U \end{array} \right\}.$$

Theorem 1.1.2 [Grothendieck (1)]. *Let X be projective over T. Then for every polynomial $P \in \mathbf{Q}[x]$ the functor $\mathcal{H}ilb^P(X/T)$ is representable by a projective T-scheme $\mathrm{Hilb}^P(X/T)$. $\mathcal{H}ilb(X/T)$ is represented by*

$$\mathrm{Hilb}(X/T) := \bigcup_{P \in \mathbf{Q}[x]} \mathrm{Hilb}^P(X/T).$$

For an open subscheme $Y \subset X$ the functor $\mathcal{H}ilb^P(Y/T)$ is represented by an open subscheme

$$\mathrm{Hilb}^P(Y/T) \subset \mathrm{Hilb}^P(X/T).$$

Definition 1.1.3. $\mathrm{Hilb}(X/T)$ is the Hilbert scheme of X over T. If T is $spec(k)$ for a field k, we will write $\mathrm{Hilb}(X)$ instead of $\mathrm{Hilb}(X/T)$ and $\mathrm{Hilb}^P(X)$ instead of $\mathrm{Hilb}^P(X/T)$. If P is the constant polynomial $P = n$, then $\mathrm{Hilb}^n(X/T)$ is the relative Hilbert scheme of subschemes of length n on X over T. If T is the spectrum of a field, we will write $X^{[n]}$ for $\mathrm{Hilb}^n(X) = \mathrm{Hilb}^n(X/spec(k))$. $X^{[n]}$ is the Hilbert scheme of subschemes of length n on X.

If U is a locally noetherian T-scheme, then $\mathcal{H}ilb^n(X/T)(U)$ is the set

$$\Big\{ \text{ closed subschemes } Z \subset X \times_T U \ \Big| \ Z \text{ is flat of degree } n \text{ over } U \Big\}.$$

In particular we can identify the set $X^{[n]}(k)$ of k-valued points of $X^{[n]}$ with the set of closed zero-dimensional subschemes of length n of X which are defined over k. In the simplest case such a subscheme is just a set of n distinct points of X with the reduced induced structure. The length of a zero-dimensional subscheme $Z \subset X$ is $dim_k H^0(Z, \mathcal{O}_Z)$. The fact that $\mathrm{Hilb}^n(X/T)$ represents the functor $\mathcal{H}ilb^n(X/T)$ means that there is a universal subscheme

$$Z_n(X/T) \subset X \times_T \mathrm{Hilb}^n(X/T),$$

which is flat of degree n over $\mathrm{Hilb}^n(X/T)$ and fulfills the following universal property: for every locally noetherian T-scheme U and every subscheme $Z \subset X \times_T U$ which is flat of degree n over U there is a unique morphism

$$f_Z : U \longrightarrow \mathrm{Hilb}^n(X/T)$$

such that

$$Z = (1_X \times_T f_Z)^{-1}(Z_n(X/T)).$$

For $T = spec(k)$ we will again write $Z_n(X)$ instead of $Z_n(X/T)$.

Remark 1.1.4. It is easy to see from the definitions that $Z_n(X/T)$ represents the functor $\mathcal{Z}_n(X/T)$ from the category of locally noetherian schemes to the category of sets which is given by

$$\mathcal{Z}_n(X/T)(U) \left\{ (Z,\sigma) \; \middle| \; \begin{array}{c} Z \text{ closed subschemes of } X \times_T U, \\ \text{flat of degree } n \text{ over } U, \\ \sigma : U \longrightarrow Z \text{ a section of the projection } Z \longrightarrow U \end{array} \right\},$$

$$\mathcal{Z}_n(X/T)(\Phi) : \mathcal{Z}_n(X/T)(U) \longrightarrow \mathcal{Z}_n(X/T)(V);$$
$$(Z,\sigma) \longmapsto (Z \times_U V, \sigma \circ \Phi)$$

(U, V locally noetherian schemes $\Phi : V \longrightarrow U$).

For the rest of section 1.1 let X be a smooth projective variety over the field k.

Definition 1.1.5. Let $G(n)$ be the symmetric group in n letters acting on X^n by permuting the factors. The geometric quotient $X^{(n)} := X^n/G(n)$ exists and is called the n-fold symmetric power of X. Let

$$\Phi_n : X^n \longrightarrow X^{(n)}$$

be the quotient map.

$X^{(n)}$ parametrizes effective zero-cycles of degree n on X, i.e. formal linear combinations $\sum n_i[x_i]$ of points x_i in X with coefficients $n_i \in \mathbb{N}$ fulfilling $\sum n_i = n$. $X^{(n)}$ has a natural stratification into locally closed subschemes:

Definition 1.1.6. Let $\nu = (n_1, \ldots, n_r)$ be a partition of n. Let

$$\Delta_{n_i} := \left\{ (x_1, \ldots, x_{n_i}) \; \middle| \; x_1 = x_2 = \ldots = x_{n_i} \right\} \subset X^{n_i}$$

be the diagonal and

$$X_\nu^n := \prod_{i=1}^r \Delta_{n_i} \subset \prod_{i=1}^r X^{n_i} = X^n.$$

Then we set

$$\overline{X_\nu^{(n)}} := \Phi_n(X_\nu^n)$$

and

$$X_\nu^{(n)} := \overline{X_\nu^{(n)}} \setminus \bigcup_{\mu > \nu} \overline{X_\mu^{(n)}}.$$

Here $\mu > \nu$ means that μ is a coarser partition then ν.

The geometric points of $X_\nu^{(n)}$ are

$$X_\nu^{(n)}(\overline{k}) = \left\{ \sum n_i[x_i] \in X^{(n)}(\overline{k}) \;\middle|\; \text{the points } x_i \text{ are pairwise distinct} \right\}.$$

The $X_\nu^{(n)}$ form a stratification of $X^{(n)}$ into locally closed subschemes, i.e they are locally closed subschemes, and every point of $X^{(n)}$ lies in a unique $X_\nu^{(n)}$. The relation between $X^{[n]}$ and $X^{(n)}$ is given by:

Theorem 1.1.7 [Mumford-Fogarty (1) 5.4]. *There is a canonical morphism (the Hilbert Chow morphism)*

$$\omega_n : X_{red}^{[n]} \longrightarrow X^{(n)},$$

which as a map of points is given by

$$Z \mapsto \sum_{x \in X} len(Z_x)[x].$$

So the above stratification of $X^{(n)}$ induces a stratification of $X_{red}^{[n]}$:

Definition 1.1.8. For every partition ν of n let

$$X_\nu^{[n]} := \omega_n^{-1}(X_\nu^{(n)}).$$

Then the $X_\nu^{[n]}$ form a stratification of $X_{red}^{[n]}$ into locally closed subschemes.

For $\nu = (n_1, \ldots, n_r)$ the geometric points of $X_\nu^{[n]}$ are just the unions of subschemes $Z_1, \ldots, Z_r$, where each Z_i is a subscheme of length n_i of X concentrated in a point x_i and the x_i are distinct.

1.2. The Weil conjectures

We will use the Weil conjectures to compute the Betti numbers of Hilbert schemes. They have been used before to compute Betti numbers of algebraic varieties, at least since in [Harder-Narasimhan (1)] they were applied for moduli spaces of vector bundles on smooth curves.

Let X be a projective scheme over a finite field $\mathbb{F}_q$, let $\overline{\mathbb{F}}_q$ be an algebraic closure of $\mathbb{F}_q$ and $\overline{X} := X \times_{\mathbb{F}_q} \overline{\mathbb{F}}_q$. The *geometric Frobenius*

$$F_X : X \longrightarrow X$$

is the morphism of X to itself which as a map of points is the identity and the map $a \mapsto a^q$ on the structure sheaf $\mathcal{O}_X$. The geometric Frobenius of $\overline{X}$ over $\mathbb{F}_q$ is

$$F_q := F_X \times 1_{\overline{\mathbb{F}}_q}.$$

The action of F_q on the geometric points $X(\overline{\mathbb{F}}_q)$ is the inverse of the action of the Frobenius of $\mathbb{F}_q$. As this is a topological generator of the Galois group $\mathrm{Gal}(\overline{\mathbb{F}}_q, \mathbb{F}_q)$, a point $x \in X(\overline{\mathbb{F}}_q)$ is defined over $\mathbb{F}_q$, if and only if $x = F_q(x)$. For a prime l which does not divide q let $H^i(\overline{X}, \mathbf{Q}_l)$ be the ith l-adic cohomology group of $\overline{X}$ and

$$b_i(\overline{X}) := dim_{\mathbf{Q}_l}(H^i(\overline{X}, \mathbf{Q}_l)),$$
$$p(\overline{X}, z) := \sum b_i(\overline{X}) z^i,$$
$$e(\overline{X}) := \sum (-1)^i b_i(\overline{X}).$$

$b_i(\overline{X})$ is independent of l. We will denote the action of F_q^* on $H^r(\overline{X}, \mathbf{Q}_l)$ by $F_q^*|_{H^r(\overline{X}, \mathbf{Q}_l)}$. The *zeta-function* of X over $\mathbb{F}_q$ is the power series

$$Z_q(X, t) := \exp\left(\sum_{n>0} |X(\mathbb{F}_{q^n})| \frac{t^n}{n} \right).$$

Here $|M|$ denotes the number of elements in a finite set M.

Let X be a smooth projective variety over the complex numbers $\mathbf{C}$. Then X is already defined over a finitely generated extension ring R of $\mathbb{Z}$, i.e. there is a variety X_R defined over R such that $X_R \times_R \mathbf{C} = X$. For every prime ideal p of R let $X_p := X_R \times_R R/p$. There is a nonempty open subset $U \subset spec(R)$ such that X_p is smooth for all $p \in U$, and the l-adic Betti-numbers of X_p coincide with those of X for all primes l different from the characteristic of A/p (cf. [Kirwan (1) 15.], [Bialynicki-Birula, Sommese (1) 2.]. If $m \subset R$ is a maximal ideal lying in U for which R/m is a finite field $\mathbb{F}_q$ of characteristic $p \neq l$, we call X_m a good reduction of X modulo q.

Theorem 1.2.1. *(Weil conjectures* [Deligne (1)], *cf.* [Milne (1)], [Mazur (1)]*)*

(1) $Z_q(X,t)$ *is a rational function*

$$Z_q(X,t) = \prod_{r=0}^{2d} Q_r(X,t)^{(-1)^{r+1}}$$

with $Q_r(X,t) = \det(1 - tF_q^*|_{H^r(\overline{X},\mathbf{Q}_l)})$.

(2) $Q_r(X,t) \in \mathbb{Z}[t]$.

(3) *The eigenvalues* $\alpha_{i,r}$ *of* $F_q^*|_{H^r(\overline{X},\mathbf{Q}_l)}$ *have the absolute value* $|\alpha_{ir}| = q^{r/2}$ *with respect to any embedding into the complex numbers.*

(4) $$Z_q(X,1/q^d t) = \pm q^{e(\overline{X})/2} t^{e(\overline{X})} Z_q(X,t).$$

(5) *If X is a good reduction of a smooth projective variety Y over $\mathbf{C}$, then we have*

$$b_i(Y) = b_i(\overline{X}) = \deg(Q_i(\overline{X},t)).$$

Remark 1.2.2. Let $F(t,s_1,\ldots,s_m) \in \mathbf{Q}[t,s_1,\ldots,s_m]$ be a polynomial. Let X and S be smooth projective varieties over $\mathbb{F}_q$ such that

$$|X(\mathbb{F}_{q^n})| = F(q^n, |S(\mathbb{F}_{q^n})|,\ldots,|S(\mathbb{F}_{q^{nm}})|)$$

holds for all $n \in \mathbb{N}$. Then we have

$$p(\overline{X}, -z) = F(z^2, p(\overline{S}, -z),\ldots, p(\overline{S}, -z^m)).$$

If X and S are good reductions of smooth varieties Y and U over $\mathbf{C}$, we have:

$$p(Y, -z) = F(z^2, p(U, -z),\ldots, p(U, -z^m)).$$

Proof: Let $\alpha_1,\ldots,\alpha_s$ be pairwise distinct complex numbers and $h_1,\ldots,h_s \in \mathbf{Q}$. We put

$$Z((\alpha_i, h_i)_i) := \exp\left(\sum_{n>0}\left(\sum_{i=1}^s h_i \alpha_i^n\right)\frac{t^n}{n}\right).$$

Then we have

$$Z((\alpha_i, h_i)_i) = \prod_{i=1}^s (1 - \alpha_i)^{-h_i}.$$

So we can read off the set of pairs $\{(\alpha_1, h_1), \dots (\alpha_s, h_s)\}$ from the function $Z((\alpha_i, h_i)_i)$. For each $c \in \mathbf{C}$ let $r(c) := 2\log_q(|c|)$. By theorem 1.2.1 we have: for a smooth projective variety W over $\mathbf{F}_q$ there are distinct complex numbers $(\beta_i)_{i=1}^t \in \mathbf{C}$ and integers $(l_i)_{i=1}^t \in \mathbf{Z}$ such that

$$|W(\mathbf{F}_{q^n})| = \sum_{i=1}^t l_i \beta_i^n$$

for all $n \in \mathbf{N}$. Furthermore we have $r(\beta_i) \in \mathbf{Z}_{\geq 0}$ and

$$(-1)^k b_k(W) = \sum_{r(\beta_i)=k} l_i$$

for all $k \in \mathbf{Z}_{\geq 0}$. Let $\beta_1, \dots, \beta_t \in \mathbf{C}$, $l_1, \dots, l_t \in \mathbf{Z}$ be the corresponding numbers for S. Then we have for all $n \in \mathbf{N}$:

$$|X(\mathbf{F}_{q^n})| = F\left(q^n, \sum_{i=1}^t l_i \beta_i^n, \dots, \sum_{i=1}^t l_i \beta_i^{mn}\right).$$

Let $\delta_1, \dots, \delta_r$ be the distinct complex numbers which appear as monomials in q and the γ_i in

$$F\left(q, \sum_{i=1}^t l_i \beta_i, \dots, \sum_{i=1}^t l_i \beta_i^m\right).$$

Then there are rational numbers $n_1, \dots, n_r$ such that

$$|X(\mathbf{F}_{q^n})| = \sum_{i=1}^r n_i \delta_i^n$$

for all $n \in \mathbf{N}$ and

$$(-1)^k b_k(\overline{X}) = \sum_{r(\delta_j)=k} n_j$$

for all $k \in \mathbf{Z}_{\geq 0}$. We see from the definitions that $\sum_{r(\delta_j)=k} n_j$ is the coefficient of z^k in $F(z^2, p(S, -z), \dots, p(S, -z^m))$. $\quad \square$

We finish by showing how to compute the number of points of the symmetric power $X^{(n)}$ for a variety X over $\mathbf{F}_q$. The geometric Frobenius $F := F_q$ acts on $X^{(n)}(\overline{\mathbf{F}}_q)$ by

$$F\left(\sum n_i [x_i]\right) = \sum n_i [F(x_i)],$$

and $X^{(n)}(\mathbf{F}_q)$ is the set of effective zero-cycles of degree n on X which are invariant under the action of F.

Definition 1.2.3. A zero-cycle of the form

$$\sum_{i=0}^{r} [F^i(x)] \qquad \text{with } x \in X(\mathbb{F}_{q^r}) \setminus \bigcup_{j|r} X(\mathbb{F}_{q^j})$$

is called a *primitive zero-cycle* of degree r on X over $\mathbb{F}_q$. The set of primitive zero-cycles of degree r on X over $\mathbb{F}_q$ will be denoted by $P_r(X, \mathbb{F}_q)$.

Remark 1.2.4.

(1) Each element $\xi \in X^{(n)}(\mathbb{F}_q)$ has a unique representation as a linear combination of distinct primitive zero-cycles over $\mathbb{F}_q$ with positive integer coefficients.

(2) $|X(\mathbb{F}_{q^n})| = \sum_{r|n} r \cdot |P_r(X, \mathbb{F}_q)|$

(3) $Z_q(X, t) = \sum_{n \geq 0} |X^{(n)}(\mathbb{F}_q)| t^n$,

i.e. $Z_q(X, t)$ is the generating function for the numbers of effective zero-cycles of X over $\mathbb{F}_q$.

Proof: (1) Let $\xi = \sum_{i=1}^{r} n_i [x_i] \in X^{(n)}(\mathbb{F}_q)$, where $x_1, \ldots, x_r$ are distinct elements of $X(\overline{\mathbb{F}}_q)$. For all j let $\xi_j := \sum_{n_i \geq j} [x_i] \in X^{(n)}(\mathbb{F}_q)$. Then we have $\xi = \sum_j \xi_j$, and it suffices to prove the result for the ξ_j. So we can assume that ξ is of the form $\xi = \sum_{i=1}^{r} [x_i]$ with pairwise distinct $x_i \in X(\overline{\mathbb{F}}_q)$. As we have $F(\xi) = \xi$, there is a permutation σ of $\{1, \ldots, r\}$ with $F(x_i) = x_{\sigma(i)}$ for all i. Let $M_1, \ldots, M_s \subset \{1, \ldots, r\}$ be the distinct orbits under the action of σ. Then we set

$$\eta_j := \sum_{i \in M_j} [x_i]$$

for $j = 1, \ldots s$. Then $\xi = \sum_{j=1}^{s} \eta_j$ is the unique representation of ξ as a sum of primitive zero-cycles.

(2) follows immediately from the definitions. From (1) we have

$$\sum_{n \geq 0} |X^{(n)}(\mathbb{F}_q)| t^n = \prod_{r \geq 1} (1 - t^r)^{-|P_r(X, \mathbb{F}_q)|}$$

$$= \exp \left(\sum_{r \geq 1} |P_r(X, \mathbb{F}_q)| \left(\sum_{m \geq 1} \frac{t^{rm}}{m} \right) \right)$$

$$= \exp \left(\sum_{n=0}^{\infty} \left(\sum_{r|n} r \cdot |P_r(X, \mathbb{F}_q)| \right) \frac{t^n}{n} \right)$$

$$= Z_q(X, t).$$

So (3) holds. $\square$

1.3. The punctual Hilbert scheme

Let $R := k[[x_1, \ldots, x_d]]$ be the field of formal power series in d variables over a field k. Let $\mathbf{m} = (x_1, \ldots, x_d)$ be the maximal ideal of R.

Definition 1.3.1. Let $I \subset R$ be an ideal of colength n. The *Hilbert function* $T(I)$ of I is the sequence $T(I) = (t_i(I))_{i \geq 0}$ of non-negative integers given by

$$t_i = dim_k(\mathbf{m}^i/(I \cap \mathbf{m}^i + \mathbf{m}^{i+1})).$$

If $T = (t_i)_{i \geq 0}$ is a sequence of non-negative integers, of which only finitely many do not vanish, we put $|T| = \sum t_i$. The *initial degree* d_0 of T is the smallest i such that $t_i < \binom{d+i-1}{i}$.

Let $R_i := \mathbf{m}^i/\mathbf{m}^{i+1}$ and $I_i := (\mathbf{m}^i \cap I)/(\mathbf{m}^{i+1} \cap I)$. Then R_i is the space of forms of degree i in R and I_i the space of initial forms of I (i.e. the forms of minimal degree among elements of I) of degree i, and we have:

$$t_i(I) = dim_k(R_i/I_i).$$

Let $I \subset R$ be an ideal of colength n and $T = (t_i)_{i \geq 0}$ the Hilbert function of I.

Lemma 1.3.2.

(1)
$$dim(\mathbf{m}^j/I \cap \mathbf{m}^j) = \sum_{i \geq j} t_i$$

holds for all $j \geq 0$. In particular we have $|T| = n$.

(2) $I \supset \mathbf{m}^n$.

Proof: Let $Z := R/I$, and Z_i the image of $\mathbf{m}^i$ under the projection $R \longrightarrow Z$. Then we have

$$\bigcap_{i \geq 0} Z_i = 0.$$

As Z is finite dimensional, there exists an i_0 with $Z_{i_0} = 0$. For such an i_0 we have $I \supset \mathbf{m}^{i_0}$. There is an isomorphism

$$Z_j = \mathbf{m}^j/(\mathbf{m}^j \cap I) \cong \oplus_{i=j}^{i_0-1} R_i/I_i$$

of k-vector spaces, and $R_i/I_i = 0$ holds for $i \geq i_0$. If we choose i_0 to be minimal, then $R_i/I_i \neq 0$ holds for $i < i_0$. So we get (1). If $t_j = 0$ for some j, then $I \supset \mathbf{m}^j$. Thus (2) follows from $|T| = n$. □

In a similar way one can prove: Let X be a smooth projective variety over an algebraically closed field k. Let $x \in X$ be a point and $Z \subset X$ a subscheme of length n with $supp(Z) = x$. Let $I_{Z,x}$ be the stalk of the ideal of Z at X. Then we have

$$I_{Z,x} \supset \mathbf{m}_{X,x}^n.$$

(Just replace R by $\mathcal{O}_{X,x}$ in the proof above.)

Remark 1.3.3. As every ideal of colength n in R contains $\mathbf{m}^n$, we can regard it as an ideal in $R/\mathbf{m}^n$. Thus the Hilbert scheme $\text{Hilb}^n(R/\mathbf{m}^n)$ also parametrizes the ideals of colength n in R. We also see that the reduced schemes $(\text{Hilb}^n(R/\mathbf{m}^k))_{red}$ are naturally isomorphic for $k \geq n$. We will therefore denote these schemes also by $\text{Hilb}^n(R)_{red}$. $\text{Hilb}^n(R)_{red}$ is the closed subscheme with the reduced induced structure of the Grassmannian $\text{Grass}(n, R/\mathbf{m}^n)$ of n dimensional quotients of $R/\mathbf{m}^n$ whose geometric points are the ideals of colength n of $\overline{k}[[x_1, \ldots, x_d]]/\mathbf{m}^n$.

Using the Hilbert function we get a stratification of $\text{Hilb}^n(R)_{red}$.

Definition 1.3.4. Let $T = (t_i)_{i \geq 0}$ be a sequence of non-negative integers with $|T| = n$. Let $Z_T \subset \text{Hilb}^n(R)_{red}$ be the locally closed subscheme (with the reduced induced structure) parametrizing ideals $I \subset R$ with Hilbert function T. Let $G_T \subset Z_T$ be the closed subscheme (with the reduced induced structure) parametrizing homogeneous ideals $I \subset R$ with Hilbert function T. Let

$$\rho_T : Z_T \longrightarrow G_T$$

be the morphism which maps an ideal I to the associated homogeneous ideal (i.e. the ideal generated by the initial forms of elements of I). The embedding $G_T \subset Z_T$ is a natural section of ρ_T.

In the case $d = 2$ i.e. $R = k[[x, y]]$ many results about these varieties have been obtained in [Iarrobino (2), (4)].

Definition 1.3.5. The jumping index $(e_i)_{i>0}$ of $(t_i)_{i \geq 0}$ is given by $e_i = max(t_{i-1} - t_i, 0)$.

Theorem 1.3.6. [Iarrobino (4), prop. 1.6, thm. 2.11, thm. 2.12, thm. 3.13]

(1) Z_T are G_T non-empty if and only if $t_0 = 1$ and $t_i \leq t_{i-1}$ for all $i \geq d_0$ (here again d_0 is the initial degree of T).

(2) G_T and Z_T are smooth, G_T is projective of dimension

$$dim(G_T) = \sum (e_j + 1)e_{j+1}.$$

(3) $\rho_T : Z_T \longrightarrow G_T$ is a locally trivial fibre bundle in the Zariski topology, whose fibre is an affine space $\mathbf{A}^{n(T)}$ of dimension

$$n(T) = n - \sum_{j \geq d_0} (e_j + 1)(e_{j+1} + e_j/2).$$

2. Computation of the Betti numbers of Hilbert schemes

The second chapter is devoted to computing the Betti numbers of Hilbert schemes of points. The main tool we want to use are the Weil conjectures. In section 2.1 we will study the structure of the closed subscheme $X_{(n)}^{[n]}$ of $X^{[n]}$ which parametrizes subschemes of length n on X concentrated in a variable point of X. We will show that $(X_{(n)}^{[n]})_{red}$ is a locally trivial fibre bundle over X in the Zariski topology with fibre $\mathrm{Hilb}^n(k[[x_1, \ldots x_d]])$. We will then also globalize the stratification of $\mathrm{Hilb}^n(k[[x_1, \ldots, x_d]])$ from section 1.3 to a stratification of $X_{(n)}^{[n]}$. Some of the strata parametrize higher order data of smooth m-dimensional subvarieties $Y \subset X$ for $m \leq d$. In chapter 3 we will study natural smooth compactifications of these strata.

In section 2.2 we consider the punctual Hilbert schemes $\mathrm{Hilb}^n(k[[x, y]])$. We give a cell decomposition of the strata and so determine their Betti numbers. I have published most of the results of this section in a different form in [Göttsche (3)]. They have afterwards been used in [Iarrobino-Yameogo (1)] to study the structure of the cohomology ring of the G_T. We also recall the results of [Ellingsrud-Strømme (1),(2)] on a cell decomposition of $\mathrm{Hilb}^n(k[[x, y]])$ and $\mathbf{P}_2^{[n]}$.

In section 2.3 we compute the Betti numbers of $S^{[n]}$ for an arbitrary smooth projective surface S using the Weil conjectures. This section gives a simplified version of my diplom paper [Göttsche (1),(2)]. The auxiliary results that we prove here will be used several times in the rest of the chapter. We also formulate a conjecture for the Hodge numbers of the $S^{[n]}$. In a joint work with Wolfgang Soergel [Göttsche-Soergel (1)] it has in the meantime been proved. Independently Cheah [Cheah (1)] has recently obtained a proof using a different method. One can see that the Euler numbers of the $S^{[n]}$ can be expressed in terms of modular forms. By the conjecture on the Hodge numbers this is also true for the signatures.

In section 2.4 we compute the Betti numbers of higher order Kummer varieties KA_n. These varieties have been defined in [Beauville (1)] as new examples of Calabi-Yau manifolds. While for a general surface S only the symmetric group $G(n)$ in n letters acts on S^n in a natural way by commuting the factors, there is also a natural action of $G(n+1)$ on A^n. KA_n can be seen as a natural desingularisation of the quotient $A^n/G(n+1)$. To determine the Betti numbers we again use the Weil conjectures. One can easily see from the formulas that the Euler numbers of the KA_n can be expressed in terms of modular forms. It was shown in [Hirzebruch-Höfer (1)] that the formula for the Euler numbers of the $S^{[n]}$ from section 2.3 coincides with the orbifold Euler number $e(S^n, G(n))$ of the action of $G(n)$. We show that the Euler number of KA_n coincides with the orbifold Euler number $e(A^n, G(n+1))$. As in section 2.3 we formulate a conjecture for the Hodge numbers. From this we also get an expression for the signatures of the KA_n in terms of modular forms.

In section 2.5 we study varieties of triangles. As mentioned above $X^{[3]}$ is smooth for an arbitrary smooth projective variety X. So we can use the Weil conjectures to compute its Betti numbers. We can view $X^{[3]}$ as a variety of unordered triangles on X. From $X^{[3]}$ we can construct several other varieties of triangles on X. The variety $\widetilde{\mathrm{Hilb}}^3(X)$ of triangles on X with a marked side has been used in [Elencwajg-Le Barz (3)] in the case of $X = \mathbf{P}_2$ to compute the Chow ring of $\mathbf{P}_2^{[3]}$, and the variety $\widehat{H}^3(X)$ of complete triangles on X has been studied in detail in [Roberts-Speiser (1),(2),(3)], [Collino-Fulton (1)] for $X = \mathbf{P}_2$. For general X it has been constructed in [Le Barz (10)]. There is also a new functorial construction by Keel [Keel (1)]. We will construct two additional varieties of triangles. We show that they are smooth and study maps and relations among the triangle varieties. Then we use the Weil conjectures to compute their Betti numbers.

2.1. The local structure of $X_{(n)}^{[n]}$

Let k be a (not necessarily algebraically closed) field and X a smooth quasiprojective variety of dimension d over k. In this section we study the structure of the stratum $(X_{(n)}^{[n]})_{red}$ which parametrizes subschemes of X which are concentrated in a (variable) point in X.

Definition 2.1.1. Let X be a smooth projective variety over a field k. Let $\Delta \subset X \times X$ be the diagonal and $\mathcal{I}_{\Delta/X \times X}$ its ideal. Let $\Delta^n \subset X \times X$ be the closed subscheme which is defined by $\mathcal{I}_{\Delta/X \times X}^n$. Let

$$p_1, p_2 : X \times X \longrightarrow X$$

be the projections and $\bar{p}_1, \bar{p}_2$ the restrictions to Δ^n. The $(n-1)^{th}$ jet-bundle $J_{n-1}(X)$ of X is the vector bundle associated to the locally free sheaf

$$\mathcal{J}_{n-1}(X) := (\bar{p}_2)_*(\mathcal{O}_{\Delta_n})$$

on X. More generally let $\mathcal{I}_{\Delta^i/\Delta^n}$ be the ideal sheaf of Δ^i in Δ^n and $J_{n-1}^i(X)$ be the vector bundle associated to

$$\mathcal{J}_{n-1}^i(X) := (\bar{p}_2)_*(\mathcal{I}_{\Delta^i/\Delta^n})$$

for all $i \leq n-1$.

We see that the fibre $J_{n-1}(X)(x)$ of $J_{n-1}(X)$ over a point $x \in X$ can be identified in a natural way with $\mathcal{O}_{X,x}/\mathbf{m}_{X,x}^n$ and similarly $J_{n-1}^i(X)(x)$ with $\mathbf{m}_{X,x}^i/\mathbf{m}_{X,x}^n$. We have

$$J_i^i(X) \cong Sym^i(T_X^*).$$

$\mathrm{Hilb}^n(\Delta^n/X)$ is a locally closed subscheme of

$$\mathrm{Hilb}^n(X \times X/X) = \mathrm{Hilb}^n(X),$$

and there is a natural morphism

$$\pi : \mathrm{Hilb}^n(\Delta^n/X) \longrightarrow X.$$

Lemma 2.1.2. $\mathrm{Hilb}^n(\Delta^n/X)_{red} = (X_{(n)}^{[n]})_{red}$ *as subschemes of* $X^{[n]}$ *and* $\pi : (X_{(n)}^{[n]})_{red} \longrightarrow X$ *is given by mapping a subscheme of length n which concentrated is in a point to this point.*

Proof: Let $\bar{k}$ be an algebraic closure of k and $\overline{X} := X \times_k \bar{k}$. Let $Z \subset X$ be a subscheme of length n of X concentrated in a point, I_Z its ideal in the local ring $\mathcal{O}_{X,x}$ and $\mathbf{m}_{X,x}$ the maximal ideal of $\mathcal{O}_{X,x}$. Then we have $I_Z \supset \mathbf{m}_{X,x}^n$ (cf. 1.3.2). So we see that $\mathrm{Hilb}^n(\Delta^n/X)_{red}$ and $(X_{(n)}^{[n]})_{red}$ are closed subschemes of $X^{[n]}$ with the reduced induced structure, which have the same geometric points. Thus they are equal. The assertion on π follows directly from the definitions. $\square$

Let $Grass(n, J_{n-1}(X))$ be the Grassmannian bundle of n-dimensional quotients of $J_{n-1}(X)$ let and $\bar{\pi} : Grass(n, J_{n-1}(X)) \longrightarrow X$ be the projection.

Lemma 2.1.3. *There is a closed embedding*

$$\bar{\imath} : \mathrm{Hilb}^n(\Delta^n/X)_{red} \longrightarrow Grass(n, J_{n-1}(X))$$

over X.

Proof: Let

$$Z_n(\Delta^n/X) \subset \Delta^n \times_X \mathrm{Hilb}^n(\Delta^n/X)$$

be the universal family (cf. 1.1.3) and let

$$\tilde{p}_2 : \Delta^n \times_X \mathrm{Hilb}^n(\Delta^n/X) \longrightarrow \mathrm{Hilb}^n(\Delta^n/X)$$

be the projection. Then we have

$$(\tilde{p}_2)_*(\mathcal{O}_{\Delta^n \times_X \mathrm{Hilb}^n(\Delta^n/X)}) = \pi^*(J_{n-1}(X)).$$

As $Z_n(\Delta^n/X)$ is flat of degree n over $\mathrm{Hilb}^n(\Delta^n/X)$, $(\tilde{p}_2)_*(\mathcal{O}_{Z_n(\Delta^n/X)})$ is a locally free quotient of rank n of $\pi^*(J_{n-1}(X))$. Thus it defines a morphism

$$i : \mathrm{Hilb}^n(\Delta^n/X) \longrightarrow Grass(n, J_{n-1}(X)).$$

So we also get a morphsim

$$\bar{\imath} : \mathrm{Hilb}^n(\Delta^n/X)_{red} \longrightarrow Grass(n, J_{n-1}(X)).$$

Let $\widetilde{T}$ be the tautological subbundle of corank n of $\bar{\pi}^*(J_{n-1}(X))$. We abreviate $Grass(n, J_{n-1}(X))$ by Y. $\bar{\pi}^*(J_{n-1}(X))$ is in a natural way an $\mathcal{O}_Y$-algebra. Let $\mathcal{Q}$ be the quotient of $\bar{\pi}^*(J_{n-1}(X))$ by the subalgebra generated by $\widetilde{T}$. $\mathcal{Q}$ is a coherent sheaf on Y. For all x in Y let

$$q(x) := dim_k(\mathcal{Q}_x \times_{\mathcal{O}_{Y,x}} \mathcal{O}_{Y,x}/\mathbf{m}_{Y,x})$$

be the rank of Q at x. From the definitions we see that $q(x) \leq n$ holds for all $x \in Y$. Let $H \subset Y$ be the closed subscheme with the reduced induced structure, for whose points $q(x) = n$ holds. Then we see

$$\bar{i}(\mathrm{Hilb}^n(\Delta^n/X)_{red}) \subset H.$$

Let $\bar{\pi} : H \longrightarrow X$ be the restriction of the projection. Let $\bar{\Delta}^n := \Delta^n \times_X H$ and $\hat{p}_2 : \bar{\Delta}^n \longrightarrow H$ be the projection. Then we have $(\hat{p}_2)_*(\mathcal{O}_{\bar{\Delta}^n}) = \bar{\pi}^*(\mathcal{J}_{n-1}(X))$. As $\hat{p}_2$ is an homeomorphism, we can view Q as a quotient of $\mathcal{O}_{\bar{\Delta}^n}$ i.e. as the structure sheaf of a subscheme Z of $\bar{\Delta}^n$, which is flat of degree n over H. This defines a morphism

$$j : H \longrightarrow \mathrm{Hilb}^n(\Delta^n/X)_{red}.$$

From the definitions it is clear that j is the inverse of i.   □

For the rest of the section we want to assume in addition that there are an open cover $(U_i)_i$ of X and local parameters on each of the U_i defined over k. Let $R := k[[x_1, \ldots, x_d]]$, $\mathbf{m} := (x_1, \ldots, x_d)$ be the maximal ideal of R and let $\mathrm{Hilb}^n(R)_{red}$ be the Hilbert scheme parametrizing ideals of colength n in $R/\mathbf{m}^n$ (cf. 1.3.3).

Lemma 2.1.4. $\pi : (X_{(n)}^{[n]})_{red} \longrightarrow X$ *is a locally trivial fibre bundle in the Zariski topology with fibre* $\mathrm{Hilb}^n(R)_{red}$.

Proof: Let $U \subset X$ be an open subset and $y_1, \ldots y_d$ local parameters on U. For each $g \in k[[x_1, \ldots, x_d]]/\mathbf{m}^n$ let

$$\tilde{g} := g((\bar{p}_2)_*(\bar{p}_1^*(y_1) - \bar{p}_2^*(y_1)), \ldots, (\bar{p}_2)_*(\bar{p}_1^*(y_d) - \bar{p}_2^*(y_d))) \in \Gamma(J_{n-1}(U)).$$

We see that the $\tilde{g}$ are a basis of $J_{n-1}(U)$ in each fibre. Thus there is an isomorphism $R/\mathbf{m}^n \otimes \mathcal{O}_U \cong J_{n-1}(U)$ and so also an isomorphism

$$\phi_n : U \times Grass(n, R/\mathbf{m}^n) \longrightarrow Grass(n, J_{n-1}(U)).$$

We see that the image of

$$U \times \mathrm{Hilb}^n(R)_{red} \subset U \times Grass(n, R/\mathbf{m}^n)$$

under ϕ_n is $\bar{\pi}^{-1}(U) = \bar{i}((U_{(n)}^{[n]})_{red})$. So the restriction of ϕ_n to $U \times \mathrm{Hilb}^n(R)_{red}$ is an isomorphism $\phi : U \times \mathrm{Hilb}^n(R)_{red} \longrightarrow (U_{(n)}^{[n]})_{red}$.   □

We can globalize the stratification of $\mathrm{Hilb}^n(R)_{red}$ to a stratification of $(X_{(n)}^{[n]})_{red}$.

Definition 2.1.5. For $i = 1, \ldots, n-1$ let

$$\psi_i : J_{n-1}^i(X) \longrightarrow J_i^i(X) \cong \operatorname{Sym}^i(T^*X)$$

be the canonical map. Let $T = (t_0, \ldots, t_{n-1})$ be a sequence of non-negative integers. Let $\bar{\pi} : Grass(n, J_{n-1}(X)) \longrightarrow X$ be the projection as above. Let $\widetilde{T}$ be the tautological subbundle of $\bar{\pi}^*(J_{n-1}(X))$. For all i let

$$Q_i := \bar{\pi}^*(J_{n-1}^i(X))/(\widetilde{T} \cap \bar{\pi}^*(J_{n-1}^i(X)) + \bar{\pi}^*(J_{n-1}^{i+1}(X))).$$

Let $W_T \subset Grass(n, J_{n-1}(X))$ be the locally closed subscheme over which the rank of Q_i is t_i for all i. Let

$$Z_T(X) = \bar{i}^{-1}(W_T) \subset (X_{(n)}^{[n]})_{red}$$

with the reduced induced structure. Let $\pi_T : Z_T(X) \longrightarrow X$ be the projection. $Q_i|_{Z_T(X)}$ is a quotient bundle of rank t_i of $\pi_T^*(\operatorname{Sym}^i(T_X^*))$.

Let T_i be the tautological subbundle on $Grass(t_i, \operatorname{Sym}^i(T_X^*))$. Let

$$\pi_1 : \prod_i Grass(t_i, \operatorname{Sym}^i(T_X^*)) \longrightarrow X$$

be the projection and

$$G_T(X) \subset \prod_i Grass(t_i, \operatorname{Sym}^i(T_X^*))$$

the closed subvariety over which

$$T_i \cdot \pi_1^*(T_X^*) \subset T_{i+1}$$

holds for all i. Here $T_1 \cdot \pi_1^*(T_X^*)$ denotes the image of $T_1 \otimes \pi_1^*(T_X^*)$ by the natural vector bundle morphism

$$\pi_1^*(\operatorname{Sym}^i(T_X^*) \otimes T_X^*) \longrightarrow \pi_1^*(\operatorname{Sym}^{i+1}(T_X^*)).$$

Let

$$\rho_T(X) : Z_T(X) \longrightarrow G_T(X)$$

be the morphism defined by the bundles $Q_i|_{Z_T(X)}$.

Analogously to the proof of lemma 2.1.4 we can easily see:

Remark 2.1.6.

(1) $Z_T(X)$ and $G_T(X)$ are locally trivial fibre bundles over X with fibres Z_T and G_T respectively.

(2) With respect to local trivialisations

$$Z_T(U) \cong U \times Z_T,$$
$$G_T(U) \cong U \times G_T$$

over an open subset $U \subset X$ we have $\rho_T(U) \cong 1_U \times \rho_T$.

Remark 2.1.7. We can see from the definitions that for all $l \le d = dim(X)$ and all $s \in \mathbb{N}$

$$G_{\left(1,l,\binom{l+1}{2},\dots,\binom{l+s-1}{s}\right)}(X) = Grass(l, T_X^*).$$

$Z_{\left(1,l,\binom{l+1}{2},\dots,\binom{l+s-1}{s}\right)}(X)$ is a locally trivial fibre bundle with fibre $\mathbf{A}^r$ over $G_{\left(1,l,\binom{l+1}{2},\dots,\binom{l+s-1}{s}\right)}(X)$. Here $r := (d-l)\left(\binom{l+s}{s} - l - 1\right)$.

Proof: By remark 2.1.6 we have to prove this only for

$$G_{\left(1,l,\binom{l+1}{2},\dots,\binom{l+s-1}{s}\right)}, Z_{\left(1,l,\binom{l+1}{2},\dots,\binom{l+s-1}{s}\right)} \subset \mathrm{Hilb}^{\binom{s+l}{s}}(R).$$

The assertion for $G_{\left(1,l,\binom{l+1}{2},\dots,\binom{l+s-1}{s}\right)}$ is obvious. Now let $Z \in Z_{\left(1,l,\binom{l+1}{2},\dots,\binom{l+s-1}{s}\right)}$ and let I_Z be the ideal of Z. Then there are $y_{l+1},\dots,y_d$ in R such that I_Z is given by $I_Z = (y_{l+1},\dots,y_d) + \mathbf{m}^{s+1}$. The initial forms u_i of the y_i all have degree 1 and are linearly independent. We can assume that $x_1,\dots,x_l,u_{l+1},\dots,u_d$ are linearly independent. We can modify the y_i to be of the form

$$y_i = u_i + f_i(x_1,\dots,x_l).$$

The $f_i(x_1,\dots,x_l)$ can be arbitrary polynomials in $x_1,\dots,x_l$ of degrees $\le s$, whose initial forms hasve degree ≥ 2. Thus the result follows. □

Remark 2.1.8. Of particular importance is the stratum $Z_{(1,\dots,1)}(X) \subset X^{[n]}_{(n)}$. It is an open subvariety of $X^{[n]}_{(n)}$. It is however in general not dense in $X^{[n]}_{(n)}$ if $d \ge 3$ and if n is large. By the definitions it parametrizes subschemes of X which are concentrated in a point x and lie on (the germ of) a smooth curve through x. We will therefore also write $X^{[n]}_{(n),c}$ instead of $Z_{(1,\dots,1)}(X)$. By remark 2.1.7 $X^{[n]}_{(n),c}$ is a locally trivial $\mathbf{A}^{(d-1)(n-2)}$-bundle over $\mathbf{P}(T_X)$.

2.2. A cell decompostion of $\mathbf{P}_2^{[n]}$, $\mathrm{Hilb}^n(R)$, Z_T, G_T

Let k be an algebraically closed field. In this section we review the methods of [Ellingsrud-Strømme(1)] for the determination of a cell decomposition of $\mathbf{P}_2^{[n]}$ and modify them in order get a cell decomposition and thus (for $k = \mathbf{C}$) the homology of the strata Z_T and G_T of $\mathrm{Hilb}^n(k[[x, y]])$. Let $R := k[[x, y]]$. Let $\mathrm{Hilb}^n(\mathbf{A}^2, 0)$ be the closed subscheme with the induced reduced structure of $(\mathbf{A}^2)^{[n]}$ parametrizing subschemes with support $\{0\}$. By lemma 2.1.4 we have

$$\mathrm{Hilb}^n(\mathbf{A}^2, 0) \cong \mathrm{Hilb}^n(R)_{red}.$$

In [Ellingsrud-Strømme (1)] the homology groups of $\mathbf{P}_2^{[n]}$, $\mathbf{A}_2^{[n]}$ and $\mathrm{Hilb}^n(\mathbf{A}^2, 0)$ are computed by constructing cell decompositions. We review some of the results and definitions on such cell decompositions. For a complex variety X let $H_*(X)$ be the Borel-Moore homology of X with $\mathbf{Z}$ coefficients. For each i let $b_i(X) = rk(H_i(X))$ be the ith Betti number and $e(X) = \sum (-1)^i b_i(X)$ the Euler number. Let $A_m(X)$ be the mth Chow group of X and $cl : A_*(X) \longrightarrow H_*(X)$ the cycle map (cf. [Fulton (1), 19.1]). For X smooth projective of dimension d we put $A^m(X) = A_{d-m}(X)$.

Definition 2.2.1. Let X be a scheme over a field k. A cell decomposition of X is a filtration

$$X = X_n \supset X_{n-1} \supset \ldots \supset X_0 \supset X_{-1} = \emptyset$$

such that $X_i \setminus X_{i-1}$ is a disjoint union of schemes $U_{i,j}$ isomorphic to affine spaces $\mathbf{A}^{n_{i,j}}$ for all $i = 0, \ldots, n$. We call the $U_{i,j}$ the cells of the decomposition.

Proposition 2.2.2. [Fulton (1) Ex. 19.1.11] *Let X be a scheme over $\mathbf{C}$ with a cell decomposition. Then*

(1) $H_{2i+1}(X) = 0$ *for all i.*

(2) $H_{2i}(X)$ *is the free abelian group generated by the homology classes of the closures of the i-dimensional cells.*

(3) *The cycle map $cl : A_*(X) \longrightarrow H_*(X)$ is an isomorphism.*

Ellingsrud and Strømme have constructed the cell decomposition of $\mathbf{P}_2^{[n]}$ using the following results of [Bialynicki-Birula (1),(2)]. Let X be a smooth projective variety over k with an action of the multiplicative group $\mathbf{G}_m$. We will denote this action by ".". Let $x \in X$ be a fixed point of this action. Let $T_{X,x}^+ \subset T_{X,x}$ be the linear subspace on which all the weights of the induced action of $\mathbf{G}_m$ are positive.

Theorem 2.2.3. [Bialynicki-Birula (1),(2)] *Let X be a smooth projective variety over an algebraically closed field k with an action of $\mathbf{G}_m$. Assume that the set of* *fixed points is the finite set* $\{x_1, \ldots, x_m\}$. *For all* $i = 1, \ldots, m$ *let*

$$X_i := \{x \in X \mid \lim_{t \to 0} t \cdot x = x_i\}.$$

Then we have:

(1) X *has a cell decomposition, whose cells are the* X_i.

(2) $T_{X_i, x_i} = T_{X, x_i}^+$.

For non-negative integers $n \geq l$ we denote by $p(n)$ the number of partitions of n and by $p(n, l)$ the number of partitions of n into l parts. This number coincides with the number of partitions of $n - l$ into numbers smaller or equal to l.

The main result of [Ellingsrud-Strømme (1)] is:

Theorem 2.2.4. [Ellingsrud-Strømme (1)]

(1) *For* $X = \mathbf{P}_2^{[n]}$, $X = \mathbf{A}_2^{[n]}$ *and* $X = \mathrm{Hilb}^n(\mathbf{A}^2, 0)$ *the following holds:* X *has a cell decomposition. In particular if* $k = \mathbf{C}$ *the cycle map* $cl : A_*(X) \longrightarrow H_*(X)$ *is an isomorphism,* $H_{2i+1}(X) = 0$, *and the* $H_{2i}(X)$ *are free abelean groups.*

If $k = \mathbf{C}$ *the Betti numbers are*

(2) $$b_{2l}(\mathbf{P}_2^{[n]}) = \sum_{n_0 + n_1 + n_2 = n} \sum_{k_0 + k_2 = l - n_1} p(n_0, n_0 - k_0) p(n_1) p(n_2, k_2 - n_2),$$

(3) $$b_{2l}(\mathbf{A}_2^{[n]}) = p(n, l - n),$$

$$b_{2l}(\mathrm{Hilb}^n(\mathbf{A}^2, 0)) = p(n, n - l).$$

We will briefly review the ideas of the proof in [Ellingsrud-Strømme (1)]. Let T_0, T_1, T_2 be a system of homogeneous coordinates on $\mathbf{P}_2$. Let $G \subset Sl(3, k)$ be the maximal torus consisting of the diagonal matrices. Let $\lambda_0, \lambda_1, \lambda_2$ be characters of G such that all the $g \in G$ can be written as

$$g = diag(\lambda_0(g), \lambda_1(g), \lambda_2(g)).$$

G acts on $\mathbf{P}_2$ by $g \cdot T_i = \lambda_i(g)T_i$. The fixed points of this action are

$$P_0 = (1, 0, 0),$$
$$P_1 = (0, 1, 0),$$
$$P_2 = (0, 0, 1).$$

The action of G on $\mathbf{P}_2$ induces an action of G on $\mathbf{P}_2^{[n]}$, as G acts on the ideals in $k[T_0, T_1, T_2]$. $Z \in \mathbf{P}_2^{[n]}$ is a fixed point if and only if the corresponding homogeneous ideal $I_Z \subset k[T_0, T_1, T_2]$ is generated by monomials. So the action on $\mathbf{P}_2^{[n]}$ has only finitely many fixed points.

Let X be a smooth projective variety over k with an action of a torus H which has only finitely many fixed points. A one-parameter subgroup $\Phi : \mathbf{G}_m \longrightarrow H$ of H which does not lie in a finite set of given hyperplanes in the lattice of one-parameter groups of H will have the same fixed points as H. In future we call such a one-parameter group "general". Thus the induced action of a general one-parameter group $\Phi : \mathbf{G}_m \longrightarrow G$ has only finitely many fixed points on $\mathbf{P}_2^{[n]}$.

Let $\Phi : \mathbf{G}_m \longrightarrow G$ be a general one-parameter group of the form $\Phi(t) = diag(t^{w_0}, t^{w_1}, t^{w_2})$ with $w_0 < w_1 < w_2$ and $w_0 + w_1 + w_2 = 0$. Let $F_0 := \{P_0\}$, $L \subset \mathbf{P}_2$ the line $T_2 = 0$, $F_1 := L \setminus P_0$, $F_2 := \mathbf{P}_2 \setminus L$. Then Φ induces the cell decomposition of $\mathbf{P}_2$ into F_0, F_1, F_2. Ellingsrud and Strømme apply theorem 2.2.3 to the induced $\mathbf{G}_m$-action on $\mathbf{P}_2^{[n]}$. We will modify their arguments in order to obtain a cell decomposition of the strata Z_T of $\mathrm{Hilb}^n(R)$.

We denote by "$\cdot$" the action of $\mathbf{G}_m$ on $\mathbf{P}_2^{[n]}$ induced by Φ. As it has only finitely many fixed points, it gives a cell decomposition of $\mathbf{P}_2^{[n]}$. $\mathrm{Hilb}^n(R)_{red} = \mathrm{Hilb}^n(\mathbf{A}^2, 0) \subset \mathbf{P}_2^{[n]}$ is the subvariety parametrizing subschemes Z of colength n with support $supp(Z) = \{P_0\}$. If $Z \in \mathbf{P}_2^{[n]}$ has support $\{P_0\}$, then

$$supp(\lim_{t \to 0}(t \cdot Z)) = \lim_{t \to 0}(t \cdot supp(Z)) = \{P_0\}.$$

If $supp(Z) \neq \{P_0\}$, then we have

$$supp(\lim_{t \to 0}(t \cdot Z)) = \lim_{t \to 0}(t \cdot supp(Z)) \neq \{P_0\}.$$

So

$$Z \in \mathrm{Hilb}^n(\mathbf{A}^2, 0) \iff \lim_{t \to 0}(t \cdot Z) \in \mathrm{Hilb}^n(\mathbf{A}^2, 0).$$

So by theorem 2.2.3 $\mathrm{Hilb}^n(\mathbf{A}^2, 0)$ is a union of cells of the cell decomposition of $\mathbf{P}_2^{[n]}$ which belong to fixed points in $\mathrm{Hilb}^n(\mathbf{A}^2, 0)$. In particular $\mathrm{Hilb}^n(\mathbf{A}^2, 0)$ has a cell decomposition.

Using the identification

$$x := T_1/T_0,$$
$$y := T_2/T_0,$$
$$R := k[[x, y]]$$

we have $\mathrm{Hilb}^n(\mathbf{A}^2, 0) = \mathrm{Hilb}^n(R)_{red}$. We identify the points of $\mathrm{Hilb}^n(R)_{red}$ with the ideals of colength n in R. The action of $\mathbf{G}_m$ on R and thus on $\mathrm{Hilb}^n(R)$ is given by

$$t \cdot x = t^{w_1 - w_0} x,$$
$$t \cdot y = t^{w_2 - w_0} y.$$

Let $I \in \text{Hilb}^n(R)$ be a fixed point. Then I is an ideal of colength n in R which is generated by monomials. Following Ellingsrud and Strømme we put

$$a_j := min\{l \mid x^j y^l \in I\}$$

for every non-negative integer j. Let r be the largest integer with $a_r > 0$. Then $(a_0, \ldots, a_r)$ is a partition of n, and $y^{a_0}, xy^{a_1}, \ldots, x^{r+1}$ are a system of generators of I. So there is a bijection between the cells of $\text{Hilb}^n(R)$ and the partitions of n. In particular the Euler number of $\text{Hilb}^n(R)$ is $p(n)$.

Let $\mathbf{T}$ be the tangent space of $\text{Hilb}^n(\mathbf{A}^2)$ in the point corresponding to I. Let Γ be a two-dimensional torus acting on R by

$$t \cdot x = \lambda(t)x,$$
$$t \cdot y = \mu(t)y$$

(here $t \in \Gamma$ and λ, μ are two linearly independent characters of Γ). We also denote by λ and μ the corresponding elements in the representation ring of Γ. By [Grothendieck (1)] there is a Γ-equivariant isomorphism

$$\mathbf{T} \cong Hom_R(I, R/I).$$

Ellingsrud and Strømme consider the corresponding representation of Γ on $\mathbf{T}$. They get:

Lemma 2.2.5. *In the representation ring of Γ there is the identity*

$$\mathbf{T} = \sum_{0 \leq i \leq j \leq r} \sum_{s=a_j+1}^{a_j-1} \left(\lambda^{i-j-1} \mu^{a_i-s-1} + \lambda^{j-i} \mu^{s-a_i} \right).$$

We give a simple proof of this result: The lemma says that $\mathbf{T}$ has a basis of common eigenvectors to Γ with the eigenvalues as in the above formula. By

$$\sum_{0 \leq i \leq j \leq r} 2(a_j - a_{j+1}) = 2 \sum_{0 \leq i \leq r} a_i = 2n = dim(\mathbf{T})$$

it is enough to give such linear independent eigenvectors. For $f \in R$ let $[f]$ be the class in R/I. An R-homomorphism $\phi : I \longrightarrow R/I$ is determined by its values on the $x^i y^{a_i}$. They must however be compatible. It is easy to see that necessary and sufficient conditions for this are

$$\phi(x^{i+1} y^{a_i}) = [x]\phi(x^i y^{a_i}),$$
$$\phi(x^i y^{a_{i-1}}) = [y^{a_{i-1}-a_i}]\phi(x^i y^{a_i}).$$

Let $0 \le i \le j \le r$ and $a_j > s \ge a_{j+1}$. Let

$$\phi_{i,j,s} : I \longrightarrow R/I; \; x^l y^{a_l} \longmapsto \begin{cases} [x^{j+l-i} y^{s+a_l-a_i}] & \text{if } l \le i, \\ 0 & \text{otherwise.} \end{cases}$$

We can see immediately that the compatibility conditions are fulfilled and $\phi_{i,j,s}$ is a common eigenvector of Γ to the eigenvalue $\lambda^{j-i} \mu^{s-a_i}$.

Let $0 \le \tilde{j} < \tilde{i} \le r+1$ and $a_{\tilde{j}} - a_{\tilde{i}-1} + a_{\tilde{i}} \le \tilde{s} < a_{\tilde{j}}$. We put

$$\phi_{\tilde{i},\tilde{j},\tilde{s}} : I \longrightarrow R/I; \; x^l y^{a_l} \longmapsto \begin{cases} [x^{\tilde{j}+l-\tilde{i}} y^{\tilde{s}+a_l-a_{\tilde{i}}}] & \text{if } l \ge \tilde{i}, \\ 0 & \text{otherwise.} \end{cases}$$

$\phi_{\tilde{i},\tilde{j},\tilde{s}}$ is an eigenvector to the eigenvalue $\lambda^{\tilde{j}-\tilde{i}} \mu^{\tilde{s}-a_{\tilde{i}}}$. The eigenvectors constructed this way are obviously linearly independent. The result follows by the substitution

$$s := \tilde{s} - a_{\tilde{j}} + a_{\tilde{i}}$$
$$j := \tilde{i} - 1$$
$$i := \tilde{j}. \qquad \square$$

We now formulate our result on the cell decompositions of Z_T and G_T in a form which has been influenced by [Iarrobino-Yameogo (1)]. In particular the formula for the Betti numbers of G_T does not follow immediately from my original formulation. In [Iarrobino-Yameogo (1)] two combinatorical formulas are shown in order to derive this formula from my original one in [Göttsche (4)]. Here we will give a direct proof.

Definition 2.2.6. Let $\alpha = (a_0, \ldots, a_r)$ be a partition of n. The graph of α is the set

$$\Gamma(\alpha) = \left\{ (i,l) \in \mathbb{Z}_{\ge 0}^2 \; \middle| \; i \le r, \; l < a_i \right\}.$$

Picturally we can represent $\Gamma(\alpha)$ as a set of points, one point in position (i,j) for each $(i,j) \in \Gamma(\alpha)$. The dual partition $\check{\alpha} = (\check{a}_1, \ldots, \check{a}_{a_0})$ is the partition, whose graph is $\Gamma(\alpha)$ with the roles of rows and columns switched. The diagonal sequence is $T(\alpha) = (t_0(\alpha), \ldots, t_l(\alpha))$, where

$$t_i(\alpha) := \left| \left\{ (l,j) \in \Gamma(\alpha) \; \middle| \; l+j = i \right\} \right|.$$

So it is the sequence of numbers of points on the diagonals of $\Gamma(\alpha)$. Let $(u,v) \in \Gamma(\alpha)$. Then the hook difference $h_{u,v}(\alpha)$ is

$$h_{u,v}(\alpha) = \left| \left\{ (u,l) \in \Gamma(\alpha) \; \middle| \; l > v \right\} \right| - \left| \left\{ (i,v) \in \Gamma(\alpha) \; \middle| \; i > u \right\} \right|.$$

I.e. $h_{u,v}(\alpha)$ is the difference of the number of points in $\Gamma(\alpha)$ in the same column above (u,v) and the number of points in the same row to the left of (u,v). So we have

$$h_{u,v}(\alpha) = \check{a}_v + v - a_u - u.$$

For the partition $\alpha = (6,3,2)$ we get for instance the diagram

$$
\begin{array}{cccccc}
\cdot & \cdot & & & & \\
\cdot & \cdot & \cdot & & & \\
\cdot & \cdot & \cdot & \cdot & \cdot & \cdot
\end{array}
$$

for $\Gamma(\alpha)$ and

$$\check{\alpha} = (3,3,2,1,1,1),$$
$$T(\alpha) = (1,2,3,3,1,1).$$

The $h_{u,v}(\alpha)$ are given by

$$
\begin{array}{cccccc}
-1 & 0 & & & & \\
-1 & 0 & 0 & & & \\
-3 & -2 & -2 & -2 & -1 & 0.
\end{array}
$$

Theorem 2.2.7. *Let $T = (t_i)_{i \geq 0}$ be a sequence of non-negative integers with $|T| = n$. Then we have for $X = G_T$ and $X = Z_T$:*

(1) *X has a cell decomposition. If $k = \mathbf{C}$, then $cl : A_*(X) \longrightarrow H_*(X)$ is an isomorphism and $H_*(X)$ is free.*

   In case $k = \mathbf{C}$ we have for the Betti numbers:

(2) $\displaystyle b_{2i}(Z_T) = \left| \left\{ \alpha \in P(n) \;\middle|\; \begin{array}{l} T(\alpha) = T; \\ |\{(u,v) \in \Gamma(\alpha) \mid h_{u,v}(\alpha) \in \{0,1\}\}| = n - i \end{array} \right\} \right|$

(3) $\displaystyle b_{2i}(G_T) = \left| \left\{ \alpha \in P(n) \;\middle|\; T(\alpha) = T; \; |\{(u,v) \in \Gamma(\alpha) \mid h_{u,v}(\alpha) = 1\}| = i \right\} \right|.$

In particular the Euler numbers are

$$e(Z_T) = e(G_T) = \left| \{ \alpha \in P(n) \mid T(\alpha) = T \} \right|.$$

Remark 2.2.8. In [Iarrobino (2),(4)] it has been shown that Z_T and G_T are non-empty if and only if $t_0 = 1$ and $t_i \leq t_{i-1}$ for all $i \geq d(T)$. If $T = (1, \ldots, 1)$, then Z_T is an $\mathbf{A}^{n-2}$-bundle over $G_T = \mathbf{P}_1$. It is easy to see that $t_0(\alpha) = 1$ and $t_i(\alpha) \leq t_{i-1}(\alpha)$ for all $i \geq d(T(\alpha))$ for each partition α of n. If $T = (1, \ldots, 1)$ the cell decomposition of Z_T of theorem 2.2.7 consists of one cell of dimension $n - 2$ and one of dimension $n - 1$ and that of G_T of one cell of dimension 0 and one cell of dimension 1 as expected.

As above let
$$\Phi : \mathbf{G}_m \longrightarrow G; \ t \mapsto diag(t^{w_0}, t^{w_1}, t^{w_2})$$

be a general one-parameter subgroup of G with $w_0 < w_1 < w_2$ and $w_0 + w_1 + w_2 = 0$. We also require the inequality

$$n(w_1 - w_0) > (n - 1)(w_2 - w_0).$$

We consider the induced $\mathbf{G}_m$-action on $\mathrm{Hilb}^n(R)$. We know already that it gives a cell decomposition of $\mathrm{Hilb}^n(R)$. Let $T = (t_i)$ be a sequence of non-negative integers with $|T| = n$.

Lemma 2.2.9.

(1) Z_T *is a union of cells of the cell decomposition of* $\mathrm{Hilb}^n(R)$.

(2) $\rho_T : Z_T \longrightarrow G_T$ *is equivariant with respect to the* $\mathbf{G}_m$*-action.*

(3) *The* $\mathbf{G}_m$*-action induces a cell decomposition of* G_T. *Its cells are the intersections of the cells of* Z_T *with* G_T.

Proof: Let I be an ideal in R with Hilbert function T. Let $j \in \mathbb{N}$, $s := j + 1 - t_j$. Let I_j be the space of initial forms of degree j in I. We put

$$J := \lim_{t \to 0} t \cdot I.$$

For all i let J_i be the space of initial forms of degree i in J. Let $T' = (t'_j)_{j \geq 0}$ be the Hilbert function of J. Choose $f_1, \ldots, f_s \in I$ such that their initial forms $g_1, \ldots, g_s$ are a basis of I_j. By replacing the f_i by suitable linear combinations we can assume that the g_i are of the form

$$g_i = x^{l(i)} y^{j-l(i)} + \sum_{m > l(i)} g_{i,m} x^m y^{j-m}$$

with $g_{i,m} \in k$ and that $l(1) > l(2) > \ldots > l(s)$. By the choice of the weights w_0, w_1, w_2 we get

$$\lim_{t \to 0} \Phi(t) \cdot (t^{l(i)(w_0 - w_1) + (j - l(i))(w_0 - w_2)} f_i) = x^{l(i)} y^{j - l(i)}.$$

So the span of the $x^{l(i)}y^{j-l(i)}$ is contained in J_j. So we have

$$t_i' = i + 1 - dim(J_i) \le t_j.$$

By $|T'| = n$ we have $T = T'$ and thus (1).

(2) follows immediately from the definitions. G_T is a smooth projective variety. If $I \in G_t$, then we have $\Phi(t) \cdot I \in G_T$ for all $t \in \mathbf{G}_m$. So $\mathbf{G}_m$ acts on G_T with a finite number of fixed points and we can apply theorem 2.2.3. As the action on G_T is the restriction of that on Z_T, (3) follows. $\square$

To determine the Betti numbers of Z_T and G_T we have to find out, which of the $\mathbf{G}_m$-invariant ideals of R lie in Z_T and what the dimensions of the corresponding cells of Z_T and G_T are. Let $\alpha = (a_0, \dots, a_r)$ be a partition of n and I the ideal of R generated by $y^{a_0}, xy^{a_1} \dots, x^{r+1}$.

Lemma 2.2.10. *For the Hilbert function $T(I)$ of I we have $T(I) = T(\alpha)$.*

Proof: Let $T(I) = (t_i)_{i \ge 0}$. The monomials $x^i y^l$ with $i + l = j$ and $l \ge a_i$ form a basis of the space I_j of homogeneous polynomials of degree j in I. So we have:

$$\begin{aligned}
t_j &= j + 1 - \left|\{(i,l) \in \mathbf{Z}_{\ge 0}^2 \mid i + l = j, \, l \ge a_i\}\right| \\
&= \left|\{(i,j) \in \Gamma(\alpha) \mid i + j = l\}\right| \\
&= t_j(\alpha). \quad \square
\end{aligned}$$

Let again $\mathbf{T}$ be the tangent space of $\mathrm{Hilb}^n(\mathbf{A}^2)$ in the point corresponding to I.

Lemma 2.2.11. *The dimension of the subspace $\mathbf{T}^+$ of $\mathbf{T}$ on which the weights of the action are positive is*

$$dim(\mathbf{T}^+) = n - \left|\{(u,v) \in \Gamma(\alpha) \mid h_{u,v}(\alpha) = 0 \text{ or } h_{u,v}(\alpha) = 1\}\right|.$$

Proof: We apply lemma 2.2.5 to $\Gamma = G$ and

$$\lambda = \frac{\lambda_1}{\lambda_0}, \ \mu = \frac{\lambda_2}{\lambda_0}.$$

Then we have for every character $\lambda^a \mu^b$ of G:

$$(\lambda^a \mu^b)(\Phi(t)) = t^{a(w_1 - w_0) + b(w_2 - w_0)}.$$

By the choice of w_0, w_1, w_2 the action of $\mathbf{G}_m$ has a positive weight on $\lambda^a \mu^b$, if and only if $a + b > 0$ or $a + b = 0$ and $b > 0$. Let i, j be integers satisfying

$$0 \leq i \leq j \leq r, \ a_{j+1} \leq s < a_j.$$

The weight of $(\lambda^{i-j-1} \mu^{a_i - s - 1}) \circ \Phi$ is positive, if and only if $i + a_i > j + s + 1$, and the weight of $\lambda^{j-i} \mu^{s-a_i}$ is positive, if and only if $i + a_i < j + s$. From the definition we see that $\breve{a}_s$ is the smallest j satisfying $s \geq a_j$, so $\breve{a}_s - 1$ is the smallest j satisfying $s \geq a_{j+1}$. So we have

$$dim\mathbf{T}^+ = \sum_{0 \leq i \leq j \leq r} \left(a_j - a_{j+1} - \left| \left\{ s \in \mathbb{Z} \ \middle| \ \begin{matrix} a_{j+1} \leq s < a_j, \\ 0 \leq j + s - i - a_i + 1 \leq 1 \end{matrix} \right\} \right| \right)$$

$$= \sum_{0 \leq i \leq r} \left(a_i - \left| \{ s \in \mathbb{Z} \mid 0 \leq s < a_i, \ 0 \leq \breve{a}_s + s - i - a_i \leq 1 \} \right| \right)$$

$$= n - \left| \{ (u, v) \in \Gamma(\alpha) \mid 0 \leq h_{u,v}(\alpha) \leq 1 \} \right|. \quad \square$$

Let $\mathbf{T}_0 \subset \mathbf{T}$ be the tangent space of G_T in I. It is easy to see that the isomorphism $\mathbf{T} \cong Hom_R(I, R/I)$ maps $\mathbf{T}_0$ to the space of degree-preserving homomorphisms in $Hom_R(I, R/I)$. In the representation ring of Γ the subspace $\mathbf{T}_0$ can be written as the sum of all terms in the representation of $\mathbf{T}$ with $a + b = 0$. Let $\mathbf{T}_0^+ \subset \mathbf{T}_0$ be the linear subspace on which the weights of the action are positive.

Lemma 2.2.12.

$$dim(\mathbf{T}_0^+) = \{ (u, v) \in \Gamma(\alpha) \mid h_{u,v} = -1 \},$$

$$dim(\mathbf{T}_0/\mathbf{T}_0^+) = \{ (u, v) \in \Gamma(\alpha) \mid h_{u,v} = 1 \}.$$

Proof: Let i, j be integers satisfying

$$0 \leq i \leq j \leq r, \ a_{j+1} \leq s \leq a_j.$$

If $i - j - 1 + a_i - s - 1 = 0$, then the weight of $(\lambda^{i-j-1} \mu^{a_i - s - 1}) \circ \Phi$ is positive. If $j - i + s - a_i = 0$, then the weight of $(\lambda^{j-i} \mu^{s-a_i})_\circ \Phi$ is negative. So we have

$$dim(\mathbf{T}_0^+) = \sum_{0 \leq i \leq j \leq r} \left| \{ s \in \mathbb{Z} \mid a_{j+1} \leq s < a_j, \ i - j + a_i - s - 2 = 0 \} \right|$$

$$= \sum_{0 \leq i \leq r} \left| \left\{ s \in \mathbb{Z} \mid 0 \leq s < a_i, \ i + a_i - s - \breve{a}_s = 1 \right\} \right|$$

$$= \left| \{ (u, v) \in \Gamma(\alpha) \mid h_{u,v}(\alpha) = -1 \} \right|$$

and

$$\dim(\mathbf{T}_0/\mathbf{T}_0^+) = \sum_{0\le i\le j\le r} |\{s\in\mathbb{Z}\mid a_{j+1}\le s < a_j,\ i+a_i-s-j=0\}|$$

$$= \sum_{0\le i\le r} \left|\left\{s\in\mathbb{Z}\ \middle|\ 0\le s < a_i,\ i+a_i-s-\check{a}_s=-1\right\}\right|$$

$$= \left|\{(u,v)\in\Gamma(\alpha)\mid h_{u,v}(\alpha)=1\}\right|. \qquad \square$$

By putting things together that theorem 2.2.7 is proved.

Remark 2.2.13. We can now easily determine the dimensions of G_T and Z_T, as they are both smooth. From lemma 2.2.12 we have:

$$\dim(G_{T(\alpha)}) = \left|\left\{(u,v)\in\Gamma(\alpha)\ \middle|\ |h_{u,v}(\alpha)|=1\right\}\right|.$$

Let $\mathbf{T}_1$ be the tangent space of $Z_{T(\alpha)}$ in I. The isomorphim $\mathbf{T}\cong Hom_R(I,R/I)$ maps $\mathbf{T}_1$ to the space of homomorphisms which preserve or increase the degree. So $\mathbf{T}_1$ can be written as the sum of the terms $\lambda^a\mu^b$ in the representation of $\mathbf{T}$ for which $a+b\ge 0$. In addition to the terms occuring in $\mathbf{T}^+$ these are exactly the $\lambda^{j-i}\mu^{s-a_i}$ with $j+s-a_i-i=0$. So we get:

$$\dim(Z_T) = \dim(\mathbf{T}_1)$$

$$= \dim(\mathbf{T}^+) + \sum_{0\le i\le j\le r}\left|\left\{s\in\mathbb{Z}\ \middle|\ \begin{array}{c} a_{j+1}\le s < a_j, \\ j+s-a_i-i+1=1 \end{array}\right\}\right|$$

$$= \dim(\mathbf{T}^+) + \left|\left\{(u,v)\in\Gamma(\alpha)\ \middle|\ h_{u,v}(\alpha)=1\right\}\right|$$

$$= n - \left|\left\{(u,v)\in\Gamma(\alpha)\ \middle|\ h_{u,v}(\alpha)=0\right\}\right|$$

Using theorem 1.3.8 we get for each partition α of n the combinatorical formulas:

$$\left|\left\{(u,v)\in\Gamma(\alpha)\ \middle|\ |h_{u,v}(\alpha)|=1\right\}\right| = \sum_{i\ge d_0(T(\alpha))} (e_i(T(\alpha))+1)e_{i+1}(T(\alpha))$$

$$\left|\left\{(u,v)\in\Gamma(\alpha)\ \middle|\ h_{u,v}(\alpha)=0\right\}\right| = \sum_{i\ge d_0(T(\alpha))} e_i(T(\alpha))(e_i(T(\alpha))+1)/2.$$

Here $(e_i(T(\alpha)))_{i>0}$ is the jumping index and $d_0(T(\alpha))$ the initial degree of $T(\alpha)$ (cf. definitions 1.3.5 and 1.3.1). In [Iarrobino-Yameogo (1)] these two formulas are proved combinatorically.

2.3. Computation of the Betti numbers of $S^{[n]}$ for a smooth surface S

We want to use the Weil conjectures to compute the Betti numbers of $S^{[n]}$ for a smooth projective surface S over $\mathbf{C}$. Let X be a smooth projective variety of dimension d over a field k. Let $R = k[[x_1, \ldots, x_d]]$. We denote $V_n := \mathrm{Hilb}^n(R)_{red}$. We denote by $len(Z_1)$ the length of a subscheme Z. For subschemes $Z_1, Z_2 \subset X$ we will write $Z_1 \subset Z_2$ if Z_1 is a subscheme of Z_2 (the same also if $Z_1 \in \mathrm{Hilb}^{n_1}(R)_{red}$ and $Z_2 \in \mathrm{Hilb}^{n_2}(R)_{red}$). For

$$(x_1, Z_1) \in (X \times V_{n_1})(\overline{k}),$$
$$(x_2, Z_2) \in (X \times V_{n_1})(\overline{k})$$

we write $(x_1, Z_1) \subset (x_2, Z_2)$, if $x_1 = x_2$ and $Z_1 \subset Z_2$. For a partition $\nu = (n_1, \ldots, n_r)$ we also write

$$\nu = (1^{\alpha_1}, 2^{\alpha_2}, \ldots),$$

where α_i is the number of summands i in ν. Let

$$|\alpha| := \sum_i \alpha_i.$$

Let $P(n)$ be the set of partitions of n.

We will assume for the following that there exist a finite open cover $(U_i)_{i=1}^s$ of X and local parameters on each of the U_i, defined over k.

Remark 2.3.1. There is a sequence of bijections $\phi_n : X_{(n)}^{[n]}(\overline{k}) \longrightarrow (X \times V_n)(\overline{k})$, commuting with the action of the Galois group $Gal(\overline{k}, k)$ such that

$$Z_1 \subset Z_2 \iff \phi_{len(Z_1)}(Z_1) \subset \phi_{len(Z_2)}(Z_2).$$

Proof: For $i = 1, \ldots, s$ let $\pi_i : ((U_i)_{(n)}^{[n]})_{red} \longrightarrow U_i$ be the restriction of the projection $\pi : (X_{(n)}^{[n]})_{red} \longrightarrow X$ from lemma 2.1.2. By lemma 2.1.4 there are isomorphisms

$$\Phi_n^i : ((U_i)_{(n)}^{[n]})_{red} \longrightarrow U_i \times V_n$$

over U_i for all $n \in \mathbf{N}$. Thus we have the required bijections $\phi_n^i := \Phi_n^i(\overline{k})$. For all $j = 1, \ldots, s$ let

$$W_j := U_j \setminus \bigcup_{i < j} U_i.$$

For each $Z \in X_{(n)}^{[n]}(\overline{k})$ there is a unique index $i(Z)$ such that $W \in \pi_{i(Z)}^{-1}(W_{i(Z)})$. We put $\phi_n(Z) := \phi_n^{i(Z)}(Z)$. The result follows, as all the ϕ_n^i are bijective. $\quad\square$

Definition 2.3.2. For any partition $\nu = (n_1, \ldots, n_r) = (1^{\alpha_1}, 2^{\alpha_2}, \ldots)$ of n let

$$\widetilde{X}_\nu^n := (\pi_{n_1} \times \ldots \times \pi_{n_r})^{-1}(X_{(1,\ldots,1)}^r) \subset X_{(n_1)}^{[n_1]} \times \ldots \times X_{(n_r)}^{[n_r]} \cong \prod_i (X_{(i)}^{[i]})^{\alpha_i}.$$

The symmetric group $G(n)$ acts on $\widetilde{X}_\nu^n$ via its quotient

$$G(\alpha) := G(\alpha_1) \times \ldots \times G(\alpha_{n_1})$$

by permuting the factors $X_{(n_i)}^{[n_i]}$ with the same n_i.

Lemma 2.3.3. *There is a natural morphism* $\psi_\nu : \widetilde{X}_\nu^n \longrightarrow X^{[n]}$, *which induces a bijection*

$$\widetilde{\psi}_\nu : \widetilde{X}_\nu^n(\overline{k})/G(n) \longrightarrow X_\nu^{[n]}(\overline{k})$$

commuting with the action of $Gal(\overline{k}, k)$.

Proof: Let T be a noetherian k-scheme and let $(Z_1, \ldots, Z_r) \in \widetilde{X}_\nu^n(T)$. We put

$$\psi_\nu((Z_1, \ldots, Z_r)) := Z_1 \cup \ldots \cup Z_r.$$

This is obviously flat of degree n over T. ψ_ν is compatible with base change, so it defines a morphism $\psi_\nu : \widetilde{X}_\nu^n \longrightarrow X^{[n]}$. The induced map $\psi_\nu(\overline{k})$ of geometric points maps $\widetilde{X}_\nu^n(\overline{k})$ to $X_\nu^{[n]}(\overline{k})$ and is invariant under the action of $G(n)$. So we have a map

$$\widetilde{\psi}_\nu : \widetilde{X}_\nu^n(\overline{k})/G(n) \longrightarrow X_\nu^{[n]}(\overline{k}).$$

The image of $Z \in X_\nu^{[n]}(\overline{k})$ is $\widetilde{\psi}_\nu^{-1}(Z) = [Z_1, \ldots, Z_r]$, where $Z_1, \ldots, Z_r$ are the connected components of Z and $[\ ]$ the class modulo $G(n)$. □

Definition 2.3.4. For an extension $\tilde{k}$ of k we write

$$V(\tilde{k}) := \bigcup_{r=0}^{\infty} V_r(\tilde{k}).$$

Let o be the point corresponding to the empty subscheme i.e. $V_0(k) = V_0(\overline{k}) = \{o\}$. For $x \in V_r(\tilde{k})$ we put $len(x) := r$. For a map $f : X(\overline{k}) \longrightarrow V(\overline{k})$ we put

$$len(f) := \sum_{x \in X(\overline{k})} len(f(x)).$$

$Gal(\overline{k}, k)$ acts on these maps by

$$\sigma(f) := \sigma \circ f \circ \sigma^{-1}.$$

We write $f_1 \subset f_2$, if $f_1(x) \subset f_2(x)$ for all $x \in X(\bar{k})$.

Lemma 2.3.5. *There exists a sequence of bijections*

$$\theta_n : X^{[n]}(\bar{k}) \longrightarrow \left\{ f : X(\bar{k}) \longrightarrow V(\bar{k}) \mid len(f) = n \right\}$$

commuting with the action of $Gal(\bar{k}/k)$ *such that*

$$Z_1 \subset Z_2 \iff \theta_{len(Z_1)}(Z_1) \subset \theta_{len(Z_2)}(Z_2).$$

Proof: Let $\nu = (n_1, \ldots, n_r)$ be a partition of n, and let $Z \in X_\nu^{[n]}(\bar{k})$ with

$$\tilde{\psi}_\nu^{-1}(Z) = [Z_1, \ldots, Z_r],$$

where $len(Z_i) = n_i$. We put

$$\theta_n(Z) := f : X(\bar{k}) \longrightarrow V(\bar{k});$$

$$x \longmapsto \begin{cases} p_2(\phi_{n_i}(Z_i)) & \text{if } x = (Z_i)_{red}, \\ o & \text{if } x \notin supp(Z), \end{cases}$$

where $p_2 : X \times V_n \longrightarrow V_n$ is the projection. The result follows from remark 2.3.1 and lemma 2.3.3. $\quad\square$

Definition 2.3.6. Now let k be a finite field $\mathbb{F}_q$, X a smooth projective variety over $\mathbb{F}_q$ and F the geometric Frobenius of X over $\mathbb{F}_q$. Let

$$P(X, \mathbb{F}_q) = \bigcup_{r>0} P_r(X, \mathbb{F}_q)$$

be the set of primitive zero cycles of X over $\mathbb{F}_q$ (cf. 1.2.3). A map $g : P(X, \mathbb{F}_q) \longrightarrow V(\bar{\mathbb{F}}_q)$ will be called *admissible*, if $g(\xi) \in V(\mathbb{F}_{q^r})$ for all $\xi \in P_r(X, \mathbb{F}_q)$. Let

$$len(g) := \sum_{\xi \in P(X, \mathbb{F}_q)} deg(\xi) len(g(\xi))$$

and

$$T_n(X, \mathbb{F}_q) := \left\{ g : P(X, \mathbb{F}_q) \longrightarrow V(\bar{\mathbb{F}}_q) \mid g \text{ admissible with } len(g) = n \right\}.$$

For $g_1 \in T_{n_1}(X, \mathbb{F}_q)$, $g_2 \in T_{n_2}(X, \mathbb{F}_q)$ we write $g_1 \subset g_2$, if $g_1(\xi) \subset g_2(\xi)$ for all $\xi \in P(X, \mathbb{F}_q)$.

Lemma 2.3.7. *There is a sequence of bijections* $\tau_n : X^{[n]} \longrightarrow T_n(X, \mathbb{F}_q)$ *such that for all subschemes* Z_1, Z_2 *of* X *of finite length*

$$\tau_{len(Z_1)}(Z_1) \subset \tau_{len(Z_2)}(Z_2) \iff Z_1 \subset Z_2$$

and such that for all $n \in \mathbb{N}$ *the following diagram commutes*

$$
\begin{array}{ccc}
X^{[n]}(\mathbb{F}_q) & \xrightarrow{\tau_n} & T_n(X, \mathbb{F}_q) \\
\downarrow{\scriptstyle \omega_n(\mathbb{F}_q)} & \swarrow{\scriptstyle \tilde{\tau}_n} & \\
X^{(n)}(\mathbb{F}_q) & &
\end{array}
$$

Here $\tilde{\tau}_n$ *is defined by*

$$\tilde{\tau}_n : T_n(X, \mathbb{F}_q) \longrightarrow X^{(n)}(\mathbb{F}_q);$$
$$g \longmapsto \sum_{\xi \in P(X, \mathbb{F}_q)} len(g(\xi)) \cdot \xi,$$

and $\omega_n : X^{[n]}_{red} \longrightarrow X^{(n)}$ *is the Hilbert-Chow morphism.*

Proof: Let

$$N_n(X, \mathbb{F}_q) := \left\{ f : M \longrightarrow V(\overline{\mathbb{F}}_q) \mid len(f) = n, F(f) = f \right\}.$$

Let

$$\tilde{\psi}_n : N_n(X, \mathbb{F}_q) \longrightarrow X^{(n)}(\mathbb{F}_q);$$
$$f \longmapsto \sum_{x \in X(\overline{\mathbb{F}}_q)} len(f(x))[x].$$

We have to find bijections $\hat{\tau}_n : N_n(X, \mathbb{F}_q) \longrightarrow T_n(X, \mathbb{F}_q)$ satisfying

$$\hat{\tau}_n(f_1) \subset \hat{\tau}_n(f_2) \iff f_1 \subset f_2,$$

such that the diagram

$$
\begin{array}{ccc}
N_n(X, \mathbb{F}_q) & \xrightarrow{\hat{\tau}_n} & T_n(X, \mathbb{F}_q) \\
\downarrow{\scriptstyle \tilde{\psi}_n(\mathbb{F}_q)} & \swarrow{\scriptstyle \tilde{\tau}_n} & \\
X^{(n)}(\mathbb{F}_q) & &
\end{array}
$$

commutes. We choose a linear ordering $\leq$ on the set $X(\overline{\mathbb{F}}_q)$. Let $f \in N_n(X, \mathbb{F}_q)$. Then $\tilde{\psi}_n(f)$ is in a unique way a linear combination

$$\tilde{\psi}_n(f) = \sum_{i=1}^{s} a_i \xi_i$$

of distinct primitive zero cycles $\xi_i \in P_{r_i}(X, \mathbb{F}_q)$ with non-negative integer coefficients a_i. For $i = 1, \ldots, s$ let $x_i \in X(\mathbb{F}_{q^{r_i}})$ be the smallest element with respect to $\le$ satisfying $\xi_i = \sum_{j=0}^{r_i - 1} [F^j(x_i)]$. Then we have

$$F^{r_i}(f(x_i)) = f(F^{r_i}(x_i)) = f(x_i),$$

so $f(x_i) \in V(\mathbb{F}_{q^{r_i}})$. We put

$$\tau_n(f) : P(X, \mathbb{F}_q) \longrightarrow V(\overline{\mathbb{F}}_q);$$
$$\xi \longmapsto \begin{cases} f(x_i) & \xi = \xi_i \text{ for a suitable } i, \\ \circ & \text{otherwise.} \end{cases}$$

The inverse τ_n^{-1} is given as follows: let $g \in T_n(X, \mathbb{F}_q)$. For $r \in \mathbb{N}$ and $\xi \in P_r(X, \mathbb{F}_q)$, let $x(\xi) \in X(\overline{\mathbb{F}}_q)$ be the smallest element $x \in X(\overline{\mathbb{F}}_q)$ with respect to $\le$ with $\xi = \sum_{j=0}^{r-1} F^j(x)$. Then we have

$$\tau_n^{-1}(g) = f : X(\overline{\mathbb{F}}_q) \longrightarrow V(\overline{\mathbb{F}}_q);$$
$$F^j(x(\xi)) \longmapsto F^j(g(\xi)). \qquad \square$$

Lemma 2.3.8.

$$\sum_{n=0}^{\infty} |X^{[n]}(\mathbb{F}_q)| t^n = \prod_{r=1}^{\infty} \left(\sum_{n=0}^{\infty} V_n(\mathbb{F}_q) t^{rn} \right)^{|P_r(X, \mathbb{F}_q)|}.$$

Proof: For all $(i, j) \in \mathbb{N} \times \mathbb{N}$ we put

$$N(i, j) := \left| \left\{ f : P_i(X, \mathbb{F}_q) \longrightarrow V(\mathbb{F}_{q^i}) \,\Big|\, \sum_{\xi \in P_i(X, \mathbb{F}_q)} len(f(\xi)) = j \right\} \right|.$$

Then by definition the number of elements of $T_n(X, \mathbb{F}_q)$ is

$$|T_n(X, \mathbb{F}_q)| = \sum_{n_1 + 2n_2 + 3n_3 + \ldots = n} \left(\prod_{s=1}^{\infty} N(s, n_s) \right).$$

On the other hand we have

$$\left(\sum_{n=0}^{\infty} |V_n(\mathbb{F}_{q^r})| t^{rn} \right)^{|P_r(X, \mathbb{F}_q)|} = \sum_{j=0}^{\infty} N(r, j) t^{rj}. \qquad \square$$

Now let S be a smooth projective surface over $\mathbb{F}_q$. Let $k := \mathbb{F}_q$, $R := k[[x, y]]$ and $V_n := \text{Hilb}^n(R)$.

Lemma 2.3.9. *For all* $l \in \mathbb{N}$ *there is an* $m_0 \in \mathbb{N}$ *such that we have for all multiples* M *of* m_0

$$\sum_{n=0}^{\infty} |S^{[n]}(\mathbb{F}_{q^M})| t^n \equiv \exp\left(\sum_{m=1}^{\infty} \frac{t^m}{m} \frac{|S(\mathbb{F}_{q^{Mm}})|}{1 - q^{Mm} t^m} \right) \quad \text{modulo } t^l.$$

Proof: Let $l \in \mathbb{N}$. There is an $m_0 \in \mathbb{N}$ such that for all $n \leq l$ the cell decomposition of $V_{n,\overline{\mathbb{F}_q}}$ from theorem 2.2.4 is already defined over $\mathbb{F}_{q^{m_0}}$. Let M be a multiple of m_0 and let $Q := q^M$. Because of the identity

$$\prod_{i=1}^{\infty} \frac{1}{1 - z^{i-1} t^i} = \sum_{n=0}^{\infty} \sum_{i=0}^{\infty} p(n, n-i) z^i t^n$$

theorem 2.2.4 implies

$$\sum_{n=0}^{\infty} |V_n(\mathbb{F}_{Q^r})| t^{rn} \equiv \prod_{i=1}^{\infty} \frac{1}{1 - Q^{r(i-1)} t^{ri}} \quad \text{modulo } t^l.$$

By lemma 2.3.8 we have:

$$\sum_{n=0}^{\infty} |S^{[n]}(\mathbb{F}_Q)| t^n = \prod_{r=1}^{\infty} \prod_{i=1}^{\infty} \left(\frac{1}{1 - Q^{r(i-1)} t^{ri}} \right)^{|P_r(S,\mathbb{F}_Q)|} \quad \text{modulo } t^l$$

$$= \exp\left(\sum_{i=1}^{\infty} \sum_{r=1}^{\infty} \sum_{h=1}^{\infty} |P_r(S,\mathbb{F}_Q)| Q^{hr(i-1)} \frac{t^{hri}}{h} \right)$$

$$= \exp\left(\sum_{i=1}^{\infty} \sum_{m=1}^{\infty} \left(\sum_{r|m} r |P_r(S,\mathbb{F}_Q)| \right) Q^{m(i-1)} \frac{t^{mi}}{m} \right)$$

$$= \exp\left(\sum_{m=1}^{\infty} \frac{t^m}{m} \frac{|S(\mathbb{F}_{Q^m})|}{1 - Q^m t^m} \right). \qquad \square$$

For the rest of this section let S be a smooth projective surface over $\mathbf{C}$. We can now compute the Poincaré polynomial

$$p(S^{[n]}, z) = \sum_{i=0}^{2n} \dim(H^i(S^{[n]}; \mathbf{Q})) z^i$$

of $S^{[n]}$. Let again $P(n)$ denote the set of partitions of n.

Theorem 2.3.10.

(1) $$p(S^{[n]}, z) = \sum_{(1^{\alpha_1}, 2^{\alpha_2}, \dots) \in P(n)} \prod_{i=1}^{\infty} p(S^{(\alpha_i)}, z) z^{2(n - |\alpha|)}$$

or equivalently:

(2)
$$\sum_{n=0}^{\infty} p(S^{[n]}, -z)t^n = \exp\left(\sum_{m=1}^{\infty} \frac{t^m}{m} \frac{p(S, -z^m)}{1 - z^{2m}t^m}\right)$$

(3)
$$\sum_{n=0}^{\infty} p(S^{[n]}, z)t^n = \prod_{m=1}^{\infty} \frac{(1 + z^{2m-1}t^m)^{b_1(S)}(1 + z^{2m+1}t^m)^{b_3(S)}}{(1 - z^{2m-2}t^m)^{b_0(S)}(1 - z^{2m}t^m)^{b_2(S)}(1 - z^{2m+2}t^m)^{b_4(S)}}$$

Proof: Let $n \in \mathbb{N}$. Let S be a smooth projective surface over $\mathbb{C}$ and S_0 a good reduction of S modulo q. Then $(S_0)^{[n]}$ is a good reduction of $S^{[n]}$ modulo q. By replacing $\mathbb{F}_q$ by a finite extension we can assume that for all $h \in \mathbb{N}$ $|(S_0)^{[n]}(\mathbb{F}_{q^h})|$ is the coefficient of t^n in

$$\exp\left(\sum_{m=1}^{\infty} \frac{t^m}{m} \frac{|S_0(\mathbb{F}_{q^{hm}})|}{1 - q^{hm}t^m}\right)$$

Now (2) follows by remark 1.2.2. (3) follows from (2) by an easy computation and (1) follows from (2) and the formula of Macdonald [Macdonald (1)]

$$\sum_{n=0}^{\infty} p(X^{[n]}, z)t^n = \prod_{i=0}^{dim_{\mathbb{R}}(X)} \left(1 + (-1)^{i+1}z^i t\right)^{(-1)^{i+1}b_i(X)}. \qquad \square$$

Corollary 2.3.11. *For the Euler numbers we have*

(1)
$$\sum_{n=0}^{\infty} e(S^{[n]})t^n = \prod_{k=1}^{\infty}(1 - t^k)^{-e(S)}$$

(2) *In particular, if $e(S) = 0$, then $e(S^{[n]}) = 0$ for all $n \in \mathbb{N}$.*

For S a two-dimensional abelian variety (2) is already known (cf. [Beauville (1), p. 769]).

Remark 2.3.12. The Euler numbers of the Hilbert schemes can be expressend in terms of modular forms: let $q := e^{2\pi i\tau}$ for τ in the upper half plane

$$\mathbf{H} := \left\{z \in \mathbb{C} \,\middle|\, \text{Im}(z) > 0\right\}.$$

Let $\Delta(\tau)$ be the cusp form of weight 12 for $Sl_2(\mathbb{Z})$ and $\eta(\tau) := \Delta(\tau)^{1/24}$ the η-function. Then

$$\sum_{n=0}^{\infty} e(S^{[n]})q^n = \left(\frac{q^{1/24}}{\eta(\tau)}\right)^{e(S)}.$$

For a K3-surface we get in particular

$$\sum_{n=0}^{\infty} e(S^{[n]})q^n = \frac{q}{\Delta(\tau)}.$$

The Betti numbers $b_i(S^{[n]})$ become stable for $n \geq i$:

Corollary 2.3.13. *Let S be a smooth irreducible surface over* **C.** *Then*

$$p(S^{[n]}, z) \equiv \prod_{m=1}^{\infty} \frac{\left((1+z^{2m-1})(1+z^{2m+1})\right)^{b_1(S)}}{(1-z^{2m})^{b_2(S)+1}(1-z^{2m+2})} \quad \text{modulo } z^{n+1}.$$

Proof: Let

$$G(z,t) := (1-t) \prod_{m=1}^{\infty} \frac{\left((1+z^{2m-1}t^m)(1+z^{2m+1}t^m)\right)^{b_1(S)}}{(1-z^{2m-2}t^m)(1-z^{2m}t^m)^{b_2(S)}(1-z^{2m+2}t^m)}.$$

We have to show

$$P(S^{[n]}, z) \equiv G(z,1) \quad \text{modulo } z^{n+1}.$$

For a power series $f \in \mathbf{Q}[[z,t]]$ we denote the coefficient of $z^i t^j$ by $a_{i,j}(f)$. We see that $a_{i,j}(G(z,t)) = 0$ holds for $i > j$. Let $i \leq n$. By theorem 2.3.10(3) we have:

$$\begin{aligned}
b_i(S^{[n]}) &= a_{i,n}\left(\left(\sum_{j=0}^{\infty} t^j\right) G(z,t)\right) \\
&= \sum_{j=0}^{n} a_{i,j}(G(z,t)) \\
&= \sum_{j=0}^{\infty} a_{i,j}(G(z,t)) \\
&= a_{i,0}(G(z,1)). \quad \square
\end{aligned}$$

The Hodge numbers of $S^{[n]}$

One would expect that similar formulas as for the Betti numbers of Hilbert schemes of points also hold for their Hodge numbers. For a smooth projective variety X over $\mathbf{C}$ let $h^{p,q}(X) := dim H^q(X, \Omega_X^p)$ be the $(p,q)^{th}$ Hodge number and let

$$h(X, x, y) := \sum_{p,q} h^{p,q}(X) x^p y^q.$$

The χ_y-genus of X is given by $\chi_y(X) = h(X, y, -1)$. By Hodge theory we have for the signature $sign(X) = \chi_1(X)$.

Together with Wolfgang Soergel I have computed the Hodge numbers of $S^{[n]}$ using intersection homology, perverse sheaves and mixed Hodge modules (cf. [Göttsche-Soergel (1)].) Independently Cheah [Cheah (1)] has recently proven this result by using a different technique, the so-called virtual Hodge polynomials. The result is:

Theorem 2.3.14.

(1) $$h(S^{[n]}, x, y) = \sum_{(1^{\alpha_1}, 2^{\alpha_2}, \ldots) \in P(n)} (xy)^{n-|\alpha|} \prod_{i=1}^{\infty} h(S^{(\alpha_i)}, x, y)$$

or equivalently

(2) $$\sum_{n=0}^{\infty} h(S^{[n]}, -x, -y) t^n = \exp \left(\sum_{m=1}^{\infty} \frac{t^m}{m} \frac{h(S, -x^m, -y^m)}{1 - (xyt)^m} \right),$$

(3) $$\sum_{n=0}^{\infty} h(S^{[n]}, x, y) t^n = \prod_{k=1}^{\infty} \prod_{p,q} \left(1 + (-1)^{p+q+1} x^{p+k-1} y^{q+k-1} t^k \right)^{(-1)^{p+q+1} h^{p,q}(S)}.$$

From this we get:

(4) $$\sum_{n=0}^{\infty} \chi_{-y}(S^{[n]}) t^n = \exp \left(\sum_{m=1}^{\infty} \frac{t^m}{m} \frac{\chi_{-y^m}(S)}{1 - (yt)^m} \right),$$

(5) $$sign(S^{[n]}) = \sum_{(1^{\alpha_1}, 2^{\alpha_2}, \ldots) \in P(n)} (-1)^{n-|\alpha|} \prod_{i=1}^{\infty} sign(S^{(\alpha_i)})$$

or equivalently

(6) $$\sum_{n=0}^{\infty} sign(S^{[n]})t^n = \prod_{k=1}^{\infty} \left(\frac{1-t^k}{1+t^k}\right)^{(-1)^k sign(S)/2} (1-t^{2k})^{-e(S)/2}.$$

(5) and (6) follow from (1) and (3) using $sign(S) = \chi_1(S)$ and $e(S) = \chi_{-1}(S)$.

Using these results we can also find formulas for the signatures of Hilbert schemes in terms of modular forms. Let again τ be in the upper half plane and $q = e^{2\pi i \tau}$. Let ϵ and δ be the following functions:

$$\epsilon := \sum_{n=1}^{\infty} \left(\sum_{d|n,\ d\ odd} \left(\frac{n}{d}\right)^3\right) q^n$$

$$\delta = -\frac{1}{8} - 3\sum_{n=1}^{\infty} \left(\sum_{d|n\ d\ odd} \frac{n}{d}\right) q^n$$

(cf. [Hirzebruch-Berger-Jung (1)], [Zagier (2)].) ϵ and δ are modular forms for $\Gamma_0(2)$ of weights 4 and 2 respectively. Both of them play an important role in the theory of elliptic genera.

Corollary 2.3.15.

$$\sum_{n=0}^{\infty} sign(S^{[n]})(-q)^n = q^{e(S)/24} \frac{\eta(\tau)^{sign(S)}}{\eta(2\tau)^{(sign(S)+e(S))/2}}$$

$$= \left(\frac{q}{\epsilon}\right)^{e(S)/24} (64(\delta^2 - \epsilon))^{sign(S)/16-e(S)/48}.$$

For a K3 surface we get in particular

$$\sum_{n=0}^{\infty} sign(S^{[n]})(-q)^n = \frac{q}{\Delta(\tau)^{2/3}\Delta(2\tau)^{1/3}}$$

$$= \frac{q}{512\epsilon(\delta^2 - \epsilon)^{3/2}}.$$

Proof: We set $t := -q$ in 2.3.14(6). Then we get

$$\sum_{n=0}^{\infty} sign(S^{[n]})(-q)^n = \prod_{k=1}^{\infty} \left(\frac{1-q^k}{1+q^k}\right)^{sign(S)/2} (1-q^{2k})^{-e(S)/2}$$

$$= \prod_{k=1}^{\infty} \frac{(1-q^k)^{sign(S)}}{(1-q^{2k})^{(sign(S)+e(S))/2}}$$

$$= q^{e(S)/24} \frac{\eta(\tau)^{sign(S)}}{\eta(2\tau)^{(sign(S)+e(S))/2}}$$

Using the formulas

$$\Delta(\tau) = 4096\epsilon(\delta^2 - \epsilon)^2$$

$$\frac{\eta(\tau)^{16}}{\eta(2\tau)^8} = 64(\delta^2 - \epsilon)$$

(cf. [Hirzebruch-Berger-Jung (1)]) we get

$$\frac{\eta(\tau)^{sign(S)}}{\eta(2\tau)^{(sign(S)+e(S))/2}} = (64(\delta^2 - \epsilon))^{sign(S)/16 - e(S)/48}\epsilon^{-e(S)/24}. \quad \square$$

2.4. The Betti numbers of higher order Kummer varieties

Definition 2.4.1. Let S be a smooth projective variety over an algebraically closed field. Let as above $\omega_n : S^{[n]} \longrightarrow S^{(n)}$ be the Hilbert-Chow morphism. Let A be the Albanese variety of S and $a : S \longrightarrow A$ be the Albanese morphism. Let $a_n : S^{(n)} \longrightarrow A^{(n)}$ be the morphism induced by a and let $g_n : A^{(n)} \longrightarrow A$ be the morphism which maps a zero-cycle $\sum [x_i]$ to its sum $\sum x_i$ in the group A. We put

$$KS_{n-1} = \omega_n^{-1}(a_n^{-1}(g_n^{-1}(0))).$$

In the following two cases we want to compute the Betti numbers of the KS_{n-1}:

(1) $S = A$ is a two-dimensional abelian variety over $\mathbf{C}$; then $a = 1_A : A \longrightarrow A$, so we have $KA_{n-1} = g_n^{-1}(0)$. In this case KA_{n-1} has been defined in [Beauville (1),(2),(3)]. There it has also been shown that KA_{n-1} is a smooth symplectic variety, and thus a new family of symplectic varieties was constructed. KA_1 is the Kummer surface of A. So we can see the KA_{n-1} as higher order Kummer varieties of A. This is the more important case.

(2) $a : S \overset{a}{\longrightarrow} A$ is a geometrically ruled surface over an elliptic curve A.

Lemma 2.4.2. *Let $S = A$ be an abelian surface, or let $a : S \longrightarrow A$ be a geometrically ruled surface over an elliptic surface A over $\mathbf{C}$ or over $\mathbb{F}_q$, where $gcd(q,n) = 1$. Then KS_{n-1} is smooth.*

Proof: For an abelian surface this has already been shown in [Beauville (1)]. We briefly repeat the argument: let $(n) : A \longrightarrow A$ be the multiplication by n. Then we have the cartesian diagram

$$
\begin{array}{ccc}
A \times KS_{n-1} & \longrightarrow & A^{[n]} \\
\downarrow & \quad \square \quad & \downarrow \\
A & \overset{(n)}{\longrightarrow} & A.
\end{array}
$$

This is true because the fibre product is

$$\{(b, Z) \in A \times A^{[n]} \mid g_n(\omega_n(Z)) = n \cdot b\},$$

and this is isomorphic to $A \times KS_{n-1}$ via $(b, Z) \mapsto (b, Z - b)$. Here $Z - b$ denotes the image of Z under the isomorphism

$$-b : A \longrightarrow A;$$
$$x \longmapsto x - b.$$

As (n) is étale, it follows that KA_{n-1} is smooth. The case of a geometrically ruled surface can be treated by a modification of this argument. Analogously to the above we have

$$
\begin{array}{ccc}
A \times K(A \times \mathbf{P}_1)_{n-1} & \longrightarrow & (A \times \mathbf{P}_1)^{[n]} \\
\downarrow & \square & \downarrow{\scriptstyle g_n \circ a_n} \\
A & \xrightarrow{\ (n)\ } & A.
\end{array}
$$

So $K(A \times \mathbf{P}_1)_{n-1}$ is smooth. Now let $S \xrightarrow{a} A$ be a geometrically ruled suface. Let $(U_i)_i$ be an open cover of A such that $a^{-1}(U_i) = U_i \times \mathbf{P}_1$ for all i. We can assume that for every effective zero-cycle ξ of length n there is an i such that $supp(\xi) \subset U_i \times \mathbf{P}_1$. Let

$$
K^i_{n-1} := \Big\{ Z \in KS_{n-1} \mid a(supp(Z)) \subset U_i \Big\}.
$$

Then the K^i_{n-1} form an open cover of KS_{n-1} with

$$
K^i_{n-1} \cong (U_i \times \mathbf{P}_1)^{[n]} \cap K(A \times \mathbf{P}_1)_{n-1}. \qquad \square
$$

We will again use the Weil conjectures to determine the Betti numbers of the KS_{n-1}. To count the points we will use a result from representation theory, the Shintani-descent. Our reference for this is [Digne (1)].

Definition 2.4.3. Let G be a group and $\langle H \rangle$ a cyclic group of automorphims of G. Let $G \ltimes \langle H \rangle$ be the semidirect product. Let j be the set-theoretic map

$$
j : G \longrightarrow G \ltimes \langle H \rangle;
$$
$$
g \longmapsto (g, H).
$$

The *H-classes* of G are the sets $j^{-1}(c)$, where c runs through the conjugacy classes of G. (G, H) has the *Lang property*, if the set of fixed points G^H is finite and each $g \in G$ can be written as $g = x^{-1} H(x)$ for an $x \in G$.

Let L be a connected algebraic group over $\mathbb{F}_q$, let $G = L(\overline{\mathbb{F}}_q)$ and F the Frobenius over $\mathbb{F}_q$. Then (G, F) has the Lang property by the theorem of Lang.

Theorem 2.4.4. *(cf. [Digne (1) Thm 1.4]). Let G be a group and H, H' two commuting automorphisms of G such that both (G, H) and (G, H') have the Lang property. Then*

(1) *For all $y \in G$ we have $y^{-1} H(y) \in G^H$.*

(2) *The map* $N_{H/H'} : y^{-1}H(y) \mapsto yH'(y^{-1})$ *definies a bijection from the set of* H-*classes of* $G^{H'}$ *to the set of* H'-*classes of* G^H.

Definition 2.4.5. Let S be a smooth projective surface over $\mathbb{F}_q$ and

$$\nu = (n_1, \ldots, n_r) = (1^{\alpha_1}, 2^{\alpha_2}, 3^{\alpha_3} \ldots)$$

a partition of n. We write as above $|\alpha| := \sum \alpha_i$, and put $|\nu| := |\alpha|$ (obviously $|\nu| = r$). As above we denote the set of partitions of n by $P(n)$. We put

$$S[\nu] := \prod_{i=1}^{\infty} \left(S^{(\alpha_i)}(\mathbb{F}_q) \times \mathbf{A}^{n-|\nu|}(\mathbb{F}_q) \right),$$

$$\gamma_\nu : S[\nu] \longrightarrow S^{(n)}(\mathbb{F}_q);$$

$$((\xi_i)_i, v) \longmapsto \sum_i i \cdot \xi_i$$

and define

$$\gamma_n : \bigcup_{\nu \in P(n)} S[\nu] \longrightarrow S^{(n)}(\mathbb{F}_q)$$

by $\gamma_n|_{S[\nu]} := \gamma_\nu$.

By theorem 2.2.4(3) we can assume (maybe after extending $\mathbb{F}_q$) that

$$|V_l(\mathbb{F}_{q^m})| = \sum_{\nu \in P(l)} q^{m(l-|\nu|)}$$

for all $l \leq n$ and all $m \in \mathbb{N}$.

Lemma 2.4.6. *For all* $\xi \in S^{(n)}(\mathbb{F}_q)$ *we have* $|\gamma_n^{-1}(\xi)| = |\omega_n^{-1}(\xi)|$.

Proof: Let

$$\xi = \sum_{i=1}^{r} n_i \xi_i \in S^{(n)}(\mathbb{F}_q),$$

where the ξ_i are distinct primitive cycles of degree d_i. Then we have

$$|\omega_n^{-1}(\xi)| = \prod_{i=1}^{r} |V_{n_i}(\mathbb{F}_{q^{d_i}})|$$

$$= \prod_{i=1}^{r} \sum_{\mu^i \in P(n_i)} q^{d_i(n_i - |\mu^i|)}$$

For $i = 1, \ldots, r$ let

$$\mu^i = (m_1^i, \ldots, m_{|\mu^i|}^i)$$

be a partition of n_i, and let

$$\nu = (n_1, \ldots n_{|\nu|})$$

be the union of d_i copies of each μ^i (i.e. if $\mu^i = (1^{\alpha_1^i}, 2^{\alpha_2^i}, \ldots)$, then $\nu = (1^{\beta_1}, 2^{\beta_2}, \ldots)$ where $\beta_j = \sum_i d_i \alpha_j^i$). Let

$$\eta_j := \sum_{i=1}^{r} \sum_{\{l \,|\, m_l^i = j\}} \xi_i.$$

Let η be the sequence $(\eta_1, \eta_2, \eta_3, \ldots)$. Then for all $w \in \mathbf{A}^{n-|\nu|}$ the pair (η, w) is an element of $S[\nu]$ and

$$\gamma_n((\eta, w)) = \xi.$$

In this way we can get all the elements of $\gamma_n^{-1}(\xi)$. So we have

$$\gamma_n^{-1}(\xi) = \sum_{\mu^1 \in P(n_1)} \sum_{\mu^2 \in P(n_2)} \cdots \sum_{\mu^r \in P(n_r)} q^{n - \sum d_i |\mu^i|}$$

$$= |\omega_n^{-1}(\xi)|. \qquad \square$$

For the next four lemmas let q be a prime power satisfying $gcd(n, q) = 1$ and let either $S = A$ be an abelian surface over $\mathbb{F}_q$ or let $S \xrightarrow{a} A$ be a geometrically ruled surface over an elliptic curve A over $\mathbb{F}_q$. In this case we assume that there exist an open cover $(U_i)_i$ of A and isomorphisms $a^{-1}(U_i) \cong U_i \times \mathbf{P}_1$ over $\mathbb{F}_q$. In both cases we assume that, for all $l \le n$, all the l-division points of A are defined over $\mathbb{F}_q$. All these conditions can be obtained by extending $\mathbb{F}_q$ if necessary. Let F be the geometric Frobenius over $\mathbb{F}_q$. We put

$$tr_l : A(\mathbb{F}_{q^l}) \longrightarrow A(\mathbb{F}_q);$$

$$x \longmapsto \sum_{i=0}^{l} F^i(x).$$

for all $l \in \mathbb{N}$.

Lemma 2.4.7. *tr_l is onto and $|tr_l^{-1}(x)|$ is independent of $x \in A(\mathbb{F}_q)$.*

Proof: We have $A(\overline{\mathbb{F}}_q)^F = (A(\mathbb{F}_q))$, and $A(\overline{\mathbb{F}}_q)^{F^l} = (A(\mathbb{F}_{q^l}))$. Let $x \in A(\mathbb{F}_{q^l})$. Choose $y \in A(\overline{\mathbb{F}}_q)$ satisfying $x = F(y) - y$ (this is possible by the Lang property). Then we have

$$N_{F^l/F} = y - F^l(y)$$

$$= \sum_{i=0}^{l-1} F^i(y - F(y))$$

$$= -tr_l(x).$$

As F^l acts as the identity on $A(\mathbb{F}_q)$, $A(\mathbb{F}_q)$ is the same as the set of F^l-classes on $A(\mathbb{F}_q)$. Thus by theorem 2.2.4 tr_l is onto. For $x \in A(\mathbb{F}_q)$ and $y \in tr_l^{-1}(x)$ the map $z \mapsto y + z$ gives a bijection between $tr_l^{-1}(0)$ and $tr_l^{-1}(x)$ $\square$

Let $h_n = g_n(\mathbb{F}_q) \circ a_n(\mathbb{F}_q) : S^{(n)}(\mathbb{F}_q) \longrightarrow A(\mathbb{F}_q)$.

Lemma 2.4.8. h_n *is onto and* $|h_n^{-1}(x)|$ *is independent of* $x \in A(\mathbb{F}_q)$.

Proof: For any partition $\nu = (n_1, \ldots, n_r) = (1^{\alpha_1}, 2^{\alpha_2}, \ldots)$ of n let $M(\nu)$ be the conjugacy class of the symmetric group $G(n)$ whose elements consist of disjoint cycles of lengths $n_1, \ldots, n_r$. Then we have

$$|M(\nu)| = \frac{n!}{\displaystyle\prod_{i=1}^{\infty} i^{\alpha_i} \alpha_i!},$$

and $\sum_{\nu \in P(n)} |M(\nu)| = n!$, as $G(n)$ is the union of the $M(\nu)$. For a smooth variety X over $\mathbb{F}_q$ we put

$$X(0, \nu) := \prod_{j=1}^{r} X(\mathbb{F}_{q^{n_j}}),$$

$$X(0, n) := \bigcup_{\nu \in P(n)} X(0, \nu),$$

$$X(n) := \bigcup_{\nu \in P(n)} X(0, \nu) \times M(\nu).$$

Let

$$\Phi_{X,0} : X(0, n) \longrightarrow X^{(n)};$$

$$(x_1, \ldots, x_r) \longmapsto \sum_{i=1}^{r} \sum_{l=0}^{n_i - 1} F^l(x_i)$$

$$\Phi_X : X(n) \longrightarrow X^{(n)};$$

$$((x_1, \ldots, x_r), m) \longmapsto \Phi_{X,0}(x_1, \ldots, x_r)$$

$$\Psi : A(n) \longrightarrow A(\mathbb{F}_q);$$

$$((a_1, \ldots, a_r), m) \longmapsto \sum_{i=1}^{r} tr_{n_i}(a_i).$$

Claim (*). $|\Phi_X^{-1}(\xi)| = n!$ for all $\xi \in X^{(n)}(\mathbb{F}_q)$.

Proof of (*): Let $\xi = \sum_{i=1}^{r} m_i\xi_i \in X^{(n)}(\mathbb{F}_q)$, where the ξ_i are distinct primitive zero-cycles of lengths d_i. The points of $\Phi_{X,0}^{-1}(\xi)$ can be obtained as follows: for any i let

$$\mu^i = (l_1^i, \dots, l_{|\mu^i|}^i) = (1^{\alpha_1^i}, 2^{\alpha_2^i}, \dots)$$

be a partition of m_i, and put $\mu := (\mu^1, \dots, \mu^r)$. Let

$$\nu(\mu) = (n_1, \dots, n_{|\nu(\mu)|}) = (1^{\beta_1}, 2^{\beta_2}, \dots)$$

be the union of the partitions $d_i \cdot \mu^i$ of $d_i m_i$ (i.e. $\beta_j = \sum_i \alpha_{j/d_i}^i$). Let

$$\rho : \bigcup_{i=1}^{r} \left(\{i\} \times \{1, \dots, |\mu^i|\} \right) \longrightarrow \{1, \dots, |\nu(\mu)|\}$$

be a bijection satisfying $n_{\rho(i,u)} = d_i l_u^i$ and $\rho(i,u) \leq \rho(i,v)$ for all $i \leq r$, $u \leq v \leq |\mu^i|$. There are

$$\frac{\prod_k \beta_k!}{\prod_{i,j} \alpha_j^i!}$$

such bijections. For all $l \in \{1, \dots, |\nu(\mu)|\}$ we choose an $x_l \in X(\mathbb{F}_{q^{n_i}})$ satisfying

$$\sum_{w=0}^{n_i-1} F^w(x_l) = l_j^i \xi_i,$$

where $\rho^{-1}(l) = (i, j)$. There are d_i choices for x_l. We see that

$$(x_1, \dots, x_{|\nu(\mu)|}) \in X(0, \nu(\mu)),$$

and we have $\Phi_{X,0}(x_1, \dots, x_{|\nu(\mu)|}) = \xi$. All the elements of $\Phi_{X,0}^{-1}(\xi)$ can be obtained this way, and all the possible choices lead to different elements. Obviously we have

$$\prod_{k=1}^{\infty} k^{\beta_k} = \prod_{i=1}^{r} \prod_{j} (jd_i)^{\alpha_j^i}.$$

Thus we get

$$\Phi_X^{-1}(\xi) = \sum_{\mu \in P(m_1) \times \ldots \times P(m_r)} \left(|\Phi_{X,0}^{-1}(\xi)| \cdot |M(\nu(\mu))| \right)$$

$$= \sum_{\mu \in P(m_1) \times \ldots \times P(m_r)} \left(\frac{n!}{\displaystyle\prod_{k=1}^{\infty} \beta_k! \, k^{\beta_k}} \cdot \frac{\displaystyle\prod_{k=1}^{\infty} \beta_k! \prod_{i=1}^{r} d_i^{|\mu^i|}}{\displaystyle\prod_{i=1}^{r} \prod_{j=1}^{\infty} \alpha_j^i!} \right)$$

$$= n! \cdot \sum_{\mu \in P(m_1) \times \ldots \times P(m_r)} \left(\frac{1}{\displaystyle\prod_{i=1}^{r} \prod_{j=1}^{\infty} (j d_i)^{\alpha_j^i}} \cdot \frac{\displaystyle\prod_{k=1}^{\infty} \prod_{i=1}^{r} d_i^{|\mu^i|}}{\displaystyle\prod_{i=1}^{r} \prod_{j=1}^{\infty} \alpha_j^i!} \right)$$

$$= n! \prod_{i=1}^{r} \sum_{\mu^i \in P(m_i)} \frac{1}{\displaystyle\prod_{j=1}^{\infty} \alpha_j^i! \, j^{\alpha_j^i}}$$

$$= n!.$$

This shows (*).

Let $\tilde{a} : S(n) \longrightarrow A(n)$ be defined by being

$$a(\mathbb{F}_{q^{n_j}}) : S(\mathbb{F}_{q^{n_j}}) \longrightarrow A(\mathbb{F}_{q^{n_j}})$$

on the factors $S(\mathbb{F}_{q^{n_j}})$ and the identity on the $M(\nu)$. The diagram

$$
\begin{array}{ccc}
S(n) & \xrightarrow{\ \Phi_S\ } & S^{(n)}(\mathbb{F}_q) \\
\Big\downarrow{\scriptstyle \tilde{a}} & & \Big\downarrow{\scriptstyle a_n(\mathbb{F}_q)} \\
A(n) & \xrightarrow{\ \Phi_A\ } & A^{(n)}(\mathbb{F}_q) \\
& \searrow{\scriptstyle \Psi} & \Big\downarrow{\scriptstyle g_n(\mathbb{F}_q)} \\
& & A(\mathbb{F}_q)
\end{array}
$$

commutes. By lemma 2.4.7. and by our assumptions before lemma 2.4.7 $|\Psi^{-1}(x)|$ and $|\tilde{a}^{-1}(\Psi^{-1}(x))|$ are independent of $x \in A(\mathbb{F}_q)$. By (*) we have

$$|h^{-1}(x)| = |\tilde{a}^{-1}(\Psi^{-1}(x))|/n!.$$

Thus the lemma follows.   □

For each $l \in \mathbb{N}$ let $A(\mathbb{F}_q)_l$ be the image of the multiplication $(l) : A(\mathbb{F}_q) \longrightarrow A(\mathbb{F}_q)$.

Lemma 2.4.9. *Let $\nu = (n_1, \ldots, n_r)$ be a partition of a number $m \in \mathbb{N}$.*

$$\sigma_\nu : A(\mathbb{F}_q)^r \longrightarrow A(\mathbb{F}_q)_{gcd(\nu)};$$

$$(x_1, \ldots, x_r) \longmapsto \sum_{i=1}^{r} n_i x_i$$

is onto and $|\sigma_\nu^{-1}(x)|$ is independent of $x \in A(\mathbb{F}_q)_{gcd(\nu)}$.

Proof: Let $x \in A(\mathbb{F}_q)_{gcd(\nu)}$ and $y \in A(\mathbb{F}_q)$ with $gcd(\nu)y = x$. Let $m_1, \ldots, m_r \in \mathbb{Z}$ satisfying

$$\sum_{i=1}^{r} m_i n_i = gcd(\nu).$$

Then we have

$$\sigma_\nu\big((m_1 y, \ldots, m_r y)\big) = x,$$

and the map

$$f_\nu : \sigma_\nu^{-1}(0) \longrightarrow \sigma_\nu^{-1}(x);$$

$$(y_1, \ldots, y_r) \longmapsto (y_1 + m_1 y, \ldots, y_r + m_r y)$$

is a bijection. $\quad \square$

Observe that

$$b_1(S) = 2dim(A) = \begin{cases} 4 & \text{in case (1) (S is an abelian surface);} \\ 2 & \text{in case (2) (A is an elliptic curve).} \end{cases}$$

Lemma 2.4.10.

(1)

$$|KS_{n-1}(\mathbb{F}_q)| = \frac{1}{|A(\mathbb{F}_q)|} \sum_{\nu = (1^{\alpha_1}, 2^{\alpha_2}, \ldots) \in P(n)} \left(gcd(\nu)^{b_1(S)} q^{n-|\nu|} \prod_{i=1}^{\infty} |S^{(\alpha_i)}(\mathbb{F}_q)| \right)$$

(2)

$$= \frac{1}{|A(\mathbb{F}_q)|} \sum_{\nu = (1^{\alpha_1}, 2^{\alpha_2}, \ldots) \in P(n)} gcd(\nu)^{b_1(S)} z^{2(n-|\nu|)}$$

$$\cdot \left(\prod_{i=1}^{\infty} \left(\sum_{\mu^i = (1^{\beta_1^i}, 2^{\beta_2^i}, \ldots) \in P(\alpha_i)} \prod_{j=1}^{\infty} \frac{1}{j^{\beta_j^i} \beta_j^i!} |S(\mathbb{F}_{q^j})|^{\beta_j^i} \right) \right)$$

Proof: By lemma 2.4.6 we have

$$|KS_{n-1}(\mathbb{F}_q)| = |\gamma_n^{-1}(h_n^{-1}(0))|$$
$$= \sum_{\nu \in P(n)} |\gamma_\nu^{-1}(h_n^{-1}(0))|.$$

Let $\nu = (n_1, \ldots, n_{|\nu|}) = (1^{\alpha_1}, 2^{\alpha_2}, \ldots)$ be a partition of n and let

$$\mu = (m_1, \ldots, m_t) := (1^{\beta_1}, 2^{\beta_2}, \ldots)$$

be defined by $\beta_i = min(1, \alpha_i)$ for all i. Let

$$f_\nu : S[\nu] \longrightarrow A(\mathbb{F}_q)^t;$$
$$((\xi_1, \ldots, \xi_t), w) \longmapsto (g_{\alpha_{m_1}}(a_{\alpha_{m_1}}(\xi_1)), \ldots, g_{\alpha_{m_t}}(a_{\alpha_{m_t}}(\xi_t))).$$

Then the diagram

$$
\begin{array}{ccc}
S[\nu] & \xrightarrow{\gamma_\nu} & S^{(n)}(\mathbb{F}_q) \\
\downarrow{\scriptstyle f_\nu} & & \downarrow{\scriptstyle h_n} \\
A(\mathbb{F}_q)^t & \xrightarrow{\sigma_\mu} & A(\mathbb{F}_q)
\end{array}
$$

commutes. By lemma 2.4.8 and lemma 2.4.9 $\sigma_\mu \circ f_\nu$ maps $S[\nu]$ onto $A(\mathbb{F}_q)_{gcd(\nu)} = A(\mathbb{F}_q)_{gcd(\mu)}$, and $|f_\nu^{-1}(\sigma_\mu^{-1}(x))|$ is is independent of $x \in A(\mathbb{F}_q)_{gcd(\nu)}$. As the multiplication with $gcd(\nu)$ is an étale morphism of degree $(gcd(\nu))^{b_1(S)}$ of A to itself, we see

$$|f_\nu^{-1}(\sigma_\nu^{-1}(0))| = \frac{|S[\nu]|}{|A(\mathbb{F}_q)_{gcd(\nu)}|}$$

$$= \frac{\left(\prod_{i=1}^{\infty} |S^{(\alpha_i)}(\mathbb{F}_q)|\right) q^{n-|\nu|}(gcd(\nu))^{b_1(S)}}{|A(\mathbb{F}_q)|}.$$

(1) follows by lemma 2.4.6, and (2) follows from this by remark 1.2.4(3) and an easy calculation. □

Theorem 2.4.11.

(1) *Let A be a two dimensional abelian variety over* **C**. *Then*

$$p(KA_{n-1}, z) = \frac{1}{(1+z)^4} \sum_{\nu=(1^{\alpha_1}, 2^{\alpha_2}, \ldots) \in P(n)} (gcd(\nu))^4 z^{2(n-|\nu|)} \prod_{i=1}^{\infty} p(A^{(\alpha_i)}, z).$$

(2) *Let S be a geometrically ruled surface over an elliptic curve over* **C**. *Then*

$$p(KS_{n-1}, z) = \frac{1}{(1+z)^2} \sum_{\nu=(1^{\alpha_1}, 2^{\alpha_2}, \ldots) \in P(n)} (gcd(\nu))^2 z^{2(n-|\nu|)} \prod_{i=1}^{\infty} p(S^{(\alpha_i)}, z).$$

(3) *In both cases we can also write these formulas as*

$$P(KS_{n-1}, -z) =$$

$$\frac{1}{(1-z)^{b_1(S)}} \sum_{\nu=(1^{\alpha_1}, 2^{\alpha_2}, \dots) \in P(n)} gcd(\nu)^{b_1(S)} z^{2(n-|\nu|)}$$

$$\cdot \left(\prod_{i=1}^{\infty} \left(\sum_{\mu^i=(1^{\beta_1^i}, 2^{\beta_2^i}, \dots) \in P(\alpha_i)} \prod_{j=1}^{\infty} \frac{1}{j^{\beta_j^i} \beta_j^i!} p(S, -z^j)^{\beta_j^i} \right) \right)$$

Proof: Let S be either a two dimensional abelian variety or a geometrically ruled surface over an elliptic curve over $\mathbf{C}$. Let $\tilde{S}$ be a good reduction of S modulo q, where $gcd(q, n) = 1$ such that the assumptions of lemma 2.4.7 hold. Then $K\tilde{S}_{n-1}$ is a good reduction of KS_{n-1} modulo q. (3) now follows by lemma 2.4.10 and remark 1.2.2. (1) and (2) follow from this by the formula of Macdonald for $p(S^{(n)}, z)$ (see the proof of theorem 2.3.10). $\quad\Box$

In section 2.3 we have obtained power series formulas for the Betti numbers of the $S^{[n]}$. We now also want to give power series for the KS_{n-1}. They will however not be as nice as those for $S^{[n]}$. We define a new multiplication $\odot$ on the ring of power series $\mathbf{Z}[[z, t, w]]$ by

$$z^{n_1} t^{m_1} w^{l_1} \odot z^{n_2} t^{m_2} w^{l_2} := z^{n_1+n_2} t^{m_1+m_2} w^{gcd(l_1, l_2)}$$

and extension by distributivity.

Proposition 2.4.12.

$$\sum_{n=0}^{\infty} p(KS_{n-1}) t^n$$

$$= \left(\frac{1}{(1+z)^{b_1(S)}} \left(w \frac{d}{dw} \right)^{b_1(S)} \right)$$

$$\bigodot_{k=1}^{\infty} \left(1 + w^k \left(-1 + \frac{(1+z^{2k-1}t^k)^{b_1(S)}(1+z^{2k+1}t^k)^{b_3(S)}}{(1-z^{2k-2}t^k)(1-z^{2k}t^k)^{b_2(S)}(1-z^{2k+2}t^k)} \right) \right) \Big|_{w=1}$$

An equivalent formula is

$$\sum_{n=0}^{\infty} p(KS_{n-1}, -z) t^n$$

$$= \left(\frac{1}{(1-z)^{b_1(S)}} \left(w \frac{d}{dw} \right)^{b_1(S)} \right)$$

$$\bigodot_{k=1}^{\infty} \left(1 + w^k \left(-1 + \exp\left(\sum_{m=1}^{\infty} p(S, z^m) \frac{z^{m(k-1)} t^m k}{m} \right) \right) \right) \Big|_{w=1}$$

Proof: It is easy to see that the two formulas are equivalent. So we only have to show the following identity:

$$\sum_{n=0}^{\infty}\left(\sum_{\nu=(1^{\alpha_1},2^{\alpha_2},\ldots)\in P(n)} w^{gcd(\nu)}\,z^{2(n-|\nu|)}\prod_{i=1}^{\infty}p(S^{(\alpha_i)},z)\right)t^n$$

$$=\bigodot_{k=1}^{\infty}\left(1+w^k\left(-1+\frac{(1+z^{2k-1}t^k)^{b_1(S)}(1+z^{2k+1}t^k)^{b_3(S)}}{(1-z^{2k-2}t^k)(1-z^{2k}t^k)^{b_2(S)}(1-z^{2k+2}t^k)}\right)\right).$$

This however follows immediately from the formula of Macdonald.   □

We can now compute the Betti numbers of the KS_{n-1} for small n. We get the following tables:

Betti numbers $b_\nu(KA_n)$ for higher order Kummer varieties:

n / ν	1	2	3	4	5	6	7	8	9	10
0	1	1	1	1	1	1	1	1	1	1
1	0	0	0	0	0	0	0	0	0	0
2	22	7	7	7	7	7	7	7	7	7
3	0	8	8	8	8	8	8	8	8	8
4	1	108	51	36	36	36	36	36	36	36
5		8	56	64	64	64	64	64	64	64
6		7	458	168	191	176	176	176	176	176
7		0	56	288	344	352	352	352	352	352
8		1	51	1046	915	786	809	794	794	794
9			8	288	1312	1528	1584	1592	1592	1592
10			7	168	3748	2879	3327	3278	3301	3286
11			0	64	1312	4496	6136	6360	6416	6424
12			1	36	915	7870	11298	12202	12571	12522
13				8	344	4496	16432	21704	23456	23680
14				7	191	2879	25524	36440	43043	44142
15				0	64	1528	16432	51640	74040	79920
16				1	36	786	11298	67049	118672	140073
17					8	352	6136	51640	162808	232368
18					7	176	3327	36440	198270	354034
19					0	64	1584	21704	162808	471712
20					1	36	809	12202	118672	538070

Betti numbers $b_\nu(KS_n)$ for S a geometrically ruled surface over an elliptic curve:

n / ν	1	2	3	4	5	6	7	8	9	10
0	1	1	1	1	1	1	1	1	1	1
1	0	0	0	0	0	0	0	0	0	0
2	6	3	3	3	3	3	3	3	3	3
3	2	4	4	4	4	4	4	4	4	4
4	6	16	13	10	10	10	10	10	10	10
5	0	8	14	16	16	16	16	16	16	16
6	1	16	45	30	35	32	32	32	32	32
7		4	32	48	54	56	56	56	56	56
8		3	45	90	108	97	102	99	99	99
9		0	14	72	142	156	162	164	164	164
10		1	13	90	247	243	278	275	280	277
11			4	48	232	348	434	448	454	456
12			3	30	247	486	668	711	738	735
13			0	16	142	472	892	1056	1146	1160
14			1	10	108	486	1206	1541	1763	1811
15				4	54	348	1232	2048	2590	2764
16				3	35	243	1206	2557	3643	4089
17				0	16	156	892	2640	4704	5824
18				1	10	97	668	2557	5737	7903
19					4	56	434	2048	5984	10028
20					3	32	278	1541	5737	11788
21					0	16	162	1056	4704	12288

Let $\sigma_1(n)$ be the sum of the positive integers dividing n. For

$$\tau \in \mathbf{H} := \left\{ a + bi \in \mathbf{C} \,\middle|\, b > 0 \right\}$$

let $q := e^{2\pi i \tau}$. Then the eta function and the Eisenstein series E_2 are given by

$$\eta(\tau) := q^{1/24} \prod_{n=1}^{\infty} (1 - q^n)$$

$$E_2(\tau) := 1 - 24 \sum_{n=1}^{\infty} \sigma_1(n) q^n.$$

We put $\tilde{\eta}(\tau) := q^{-1/24}\eta(\tau)$.

Corollary 2.4.13.

(1) *For an abelian surface A over $\mathbf{C}$ we have*

$$e(KA_{n-1}) = n^3 \sigma_1(n).$$

(2) *For a geometrically ruled surface over an elliptic curve we have*

$$e(KS_{n-1}) = 2n\sigma_1(n).$$

(3) *In both cases this can be expressed in terms of modular forms as*

$$\sum_{n=1}^{\infty} e(KS_{n-1})q^n = \frac{3 - b_1(S)/2}{4\pi i} \left(q\frac{d}{dq}\right)^{b_1(S)-1} \frac{d}{d\tau} \log(\tilde{\eta}(\tau))$$

$$= \frac{1}{24}(3 - b_1(S)/2) \left(q\frac{d}{dq}\right)^{b_1(S)-1} E_2.$$

Proof: As $p(S, -z^i)$ is divisible by $(1 - z)^{b_1(S)}$, we see that every summand

$$\frac{1}{(1 - z)^{b_1(S)}} \prod_{i=1}^{\infty} \prod_{j=1}^{\infty} \frac{1}{j^{\beta_j^i} \beta_j^i!} p(S, -z^j)^{\beta_j^i}$$

in the sum of theorem 2.4.11(3) is divisible by

$$(1 - z)^{b_1(S)((\sum_{i,j} \beta_j^i)-1)}.$$

Thus it does not contribute to the Euler number, except if ν is $(n_1^{\frac{n}{n_1}})$ and μ^{n_1} is $(\frac{n}{n_1})$ for some divisor n_1 of n. So we get from theorem 2.4.11(3):

$$e(KS_{n-1}) = \sum_{n_1 | n} n_1^{b_1(S)} \frac{n_1}{n} \frac{p(S, -z^{n/n_1})}{(1 - z)^{b_1(S)}} \bigg|_{z=1}$$

$$= \sum_{n_1 | n} n_1^{b_1(S)} \frac{n_1}{n} (3 - b_1(S)/2) \left(\frac{n}{n_1}\right)^{b_1(S)}$$

$$= (3 - b_1(S)/2)n^{b_1(S)-1}\sigma_1(n). \qquad \square$$

Table of the Euler numbers (A abelian surface, S geometrically ruled surface over an elliptic curve):

n	1	2	3	4	5	6	7	8	9	10
ν										
$e(KA_n)$	24	108	448	750	2592	2744	7680	9744	18000	15972
$e(KS_n)$	12	24	56	60	144	112	240	234	360	264

We can again see easily that the Betti numbers $b_i(KS_{n-1})$ become stable for $i \leq n$.

Corollary 2.4.14.

$$p(KS_{n-1}, z) \equiv (1 - z^2) \prod_{m=1}^{\infty} \left(\frac{1 + z^{2m+1}}{1 - z^{2m}} \right)^{2b_1(S)} \quad modulo\ z^n.$$

Proof: For any partition ν of n satisfying $|\nu| > n/2$ we see that $gcd(\nu) = 1$. So we have

$$p(KS_{n-1}, z) \equiv \frac{p(S^{[n]}, z)}{(1 + z)^{b_1(S)}} \quad modulo\ z^n.$$

Thus we have by corollary 2.3.13

$$p(KS_{n-1}, z) \equiv \frac{1}{(1 + z)^{b_1(S)}} \prod_{m=1}^{\infty} \frac{(1 + z^{2m-1})^{b_1(S)}(1 + z^{2m+1})^{b_1(S)}}{(1 - z^{2m})^{b_2(S)+1}(1 - z^{2m+2})} \quad modulo\ z^n$$

$$= \prod_{m=1}^{\infty} \frac{(1 + z^{2m+1})^{2b_1(S)}}{(1 - z^{2m})^{b_2(S)+1}(1 - z^{2m+2})}$$

$$= (1 - z^2) \prod_{m=1}^{\infty} \frac{(1 + z^{2m+1})^{2b_1(S)}}{(1 - z^{2m})^{b_2(S)+2}}$$

The result follows. □

In particular we have $b_1(KS_{n-1}) = 0$ for all $n \in \mathbb{N}$. In fact the KA_{n-1} were proven to be simply connected in [Beauville (1)].

The orbifold Euler number formula

Let G be a finite group acting on a compact differentiable manifold X. Then there exists the well known formula for the Euler number of the quotient

$$e(X/G) = \frac{1}{|G|} \sum_{g \in G} e(X^g),$$

where X^g denotes the set of fixed points of $g \in G$. If the quotient X/G is viewed as an orbifold, it still carries information on the action of G. In [Dixon-Harvey-Vafa-Witten (1),(2)] the orbifold Euler number is defined by

$$e(X, G) = \frac{1}{|G|} \sum_{gh = hg} e(X^g \cap X^h)$$

(the sum is over all commuting pairs of elements in G). Now let X be an algebraic variety. We assume that the canonical divisor $K_{X/G}$ of X/G exists as a Cartier divisor. Furthermore we assume that there is a resolution $\widehat{X/G} \xrightarrow{\pi} X/G$ satisfying $K_{\widehat{X/G}} = \pi^* K_{X/G}$. Then it has been conjectured that

$$e(X, G) = e(\widehat{X/G}).$$

This formula we will call the orbifold Euler number formula. In the case that the group G is abelian this conjecture has been proved in [Roan (1)] under certain additional hypotheses. In [Hirzebruch-Höfer (1)] some examples of this formula are studied. First they give a reformulation:

$$e(X, G) = \sum_{[g]} e(X^g / C(g)).$$

Here $C(g)$ is the centralizer of g and $[g]$ runs through the conjugacy classes of G. Hirzebruch and Höfer consider in particular the action of the symmetric group $G(n)$ on the nth power S^n of a smooth projective surface S by permuting the factors. The quotient is the symmetric power $S^{(n)}$, and $\omega_n : S^{[n]} \longrightarrow S^{(n)}$ is a canonical resolution of $S^{(n)}$. The canonical divisor K_{S^n} is invariant under the $G(n)$ action. Thus it gives a canionical Cartier divisor $K_{S^{(n)}}$ on $S^{(n)}$, and it is easy to show that

$$\omega_n^*(K_{S^{(n)}}) = K_{S^{[n]}}.$$

So the assumptions of the conjecture are fulfilled, and in fact Hirzebruch and Höfer use my formulas (corollary 2.3.11) to prove that

$$e(S^{[n]}) = e(S^n, G(n)).$$

Another case in which they check the formula is that of the Kummer surface KA_1 of an abelian surface as a resolution of the quotient of A by $G(2) = \mathbb{Z}/2$ acting by $x \mapsto -x$. We will now generalize this result to the higher order Kummer varieties KA_{n-1}. Let A be an abelian surface. Let

$$A_0^{(n)} := g_n^{-1}(0) = \left\{ \sum [x_i] \in A^{(n)} \ \Big| \ \sum x_i = 0 \right\} \subset A^{(n)},$$
$$A_0^n := \left\{ (x_1, \ldots, x_n) \in A^n \ \Big| \ \sum x_i = 0 \right\} \subset A^n$$

with the reduced induced structure. Then A_0^n is isomorphic to A^{n-1}. The $G(n)$ action by permutation of the factors of A^n maps A_0^n to itself. So we can restrict it to A_0^n and the quotient is $A_0^{(n)}$. Let $\omega := \omega_n|_{KA_{n-1}}$. Then $\omega : KA_{n-1} \longrightarrow A_0^{(n)}$ is a canonical desingularisation of $A_0^{(n)}$. The canonical divisor of $A_0^{(n)}$ is trivial, and by [Beauville (1)] KA_{n-1} is a symplectic variety; in particular we also have $K_{KA_{n-1}} = 0$. So the conjecture says that $e(KA_{n-1}) = e(A^{n-1}, G(n))$ should hold.

For a permutation σ of $\{1, \ldots, n\}$ let

$$p(\sigma) = (1^{\alpha_1(\sigma)}, 2^{\alpha_2(\sigma)}, \ldots)$$

be the partition of n which consists of the lengths of the cycles of σ. It determines the conjugacy class of σ. The fixed point set is given by

$$(A^n)^{\sigma} = \left\{ (x_1, \ldots, x_n) \in A^n \ \Big| \ x_{\nu_1} = \ldots = x_{\nu_i} \text{ for all cycles } (\nu_1, \ldots, \nu_i) \text{ of } \sigma \right\}$$
$$\cong \prod_{i=1}^{\infty} A^{\alpha_i(\sigma)}.$$

The centralizer $C(\sigma)$ acts by permuting the cycles of σ of the same lengths. So we get

$$(A^n)^{\sigma}/C(\sigma) \cong \prod_i A^{(\alpha_i(\sigma))}$$
$$= \prod_i A^{\alpha_i(\sigma)}/G(\alpha_i(\sigma)).$$

For

$$h = (h_1, h_2, \ldots) \in \prod_i G(\alpha_i(\sigma)),$$

the fixed point set $((A^n)^{\sigma})^h$ consists of the $(x_1, \ldots, x_n) \in A^n$ satisfying $x_i = x_j$ for all i, j for which the following holds: either i and j occur in the same cycle of σ, or they occur in two different cycles of the same length l, and these are permuted by h_l. So we get that $((A^n)^{\sigma})^h = (A^n)^{\tau}$ for some $\tau \in G(n)$ and

$$((A^n)^{\sigma})^h = (A^n)^{(1, \ldots, n)} \cong A,$$

if and only if
$$p(\sigma) = ((n/\alpha)^{\alpha}),$$
and h_{α} is a cycle of length α in $G(\alpha)$ for a positive integer α dividing n.

Remark 2.4.15. Let $\sigma \in G(n)$. Then we have
$$e((A_0^n)^{\sigma}) = \begin{cases} n^4 & p(\sigma) = (n); \\ 0 & \text{otherwise.} \end{cases}$$

Proof: Let B be an abelian variety and $h : B \longrightarrow B$ an automorphism of B. Then every connected component of B^h is either an isolated point or a translation of an abelian subvariety of positive dimension of B. In particular $e(B^h)$ is the number of isolated points in B^h. For a cycle σ of length n we have
$$(A_0^n)^{\sigma} = \left\{ (x_1, \ldots, x_n) \;\middle|\; x_1 = \ldots = x_n, \sum x_i = 0 \right\}$$
$$\cong \left\{ x \in A \;\middle|\; nx = 0 \right\},$$
and this has Euler number n^4. Let $\sigma \in G(n)$ with
$$p(\sigma) = (n_1, \ldots, n_r), \quad r \geq 2.$$
Then we get
$$(A_0^n)^{\sigma} \cong \left\{ (x_1, \ldots, x_r) \;\middle|\; \sum n_i x_i = 0 \right\}.$$
Let $(x_1, \ldots, x_r) \in (A_0^n)^{\sigma}$. For every $y \in A$ the point
$$(x_1 + n_2 y, x_2 - n_1 y, x_3, \ldots, x_r)$$
lies in the same connected component of $(A_0^n)^{\sigma}$ as $(x_1, \ldots, x_r)$. By the above we have $e((A_0^n)^{\sigma}) = 0$. □

Theorem 2.4.16. $e(A^{n-1}, G(n)) = n^3 \sigma_1(n) = e(KA_{n-1})$.

Proof:
$$e(A_0^n, G(n)) = \sum_{[\sigma]} e((A_0^n)^{\sigma}/C(\sigma))$$
$$= \sum_{[\sigma]} e\left((A_0^n)^{\sigma} \middle/ \left(\prod_i G(\alpha_i(\sigma)) \right) \right)$$
$$= \sum_{[\sigma]} \sum_{h \in \prod_i G(\alpha_i(\sigma))} \frac{1}{\prod_i \alpha_i(\sigma)!} e(((A_0^n)^{\sigma})^h)$$
$$= \sum_{\alpha|n} \frac{n^4}{\alpha}$$
$$= n^3 \sigma_1(n) \qquad □$$

Conjectures on the Hodge numbers of the KS_{n-1}

Similar to the results of theorem 2.3.14 we can formulate conjectures on the Hodge numbers of the KS_{n-1}.

Conjecture 2.4.17.

$$
h(KS_{n-1}, x, y)
$$
$$
= \frac{1}{((1+x)(1+y))^{b_1(S)/2}} \sum_{\nu=(1^{\alpha_1}, 2^{\alpha_2}, \dots) \in P(n)} (gcd(\nu))^{b_1(S)}(xy)^{n-|\nu|}
$$
$$
\cdot \prod_i h(S^{(\alpha_i)}, x, y)
$$

or equivalently

$$
h(KS_{n-1}, -x, -y)
$$
$$
= \frac{1}{(1-x)^{b_1(S)/2}(1-y)^{b_1(S)/2}} \sum_{\nu=(1^{\alpha_1}, 2^{\alpha_2}, \dots) \in P(n)} gcd(\nu)^{b_1(S)}(xy)^{n-|\nu|}
$$
$$
\cdot \prod_i \left(\sum_{\mu^i = (1^{\beta_1^i}, 2^{\beta_2}, \dots) \in P(\alpha_i)} \prod_j \frac{1}{j^{\beta_j^i} \beta_j^i!} h(S, -x^j, -y^j)^{\beta_j^i} \right).
$$

In the case of the KA_{n-1} the conjecture has been verified in [Göttsche-Soergel (1)].

Remark 2.4.18. From the proven part of conjecture 2.4.17 we get for the χ_y-genus and the signature:

(1)
$$
\chi_{-y}(KA_{n-1}) = n \sum_{n_1 | n} n_1^3 (1 + y \dots + y^{n/n_1 - 1})^2 y^{n - n/n_1},
$$

(2)
$$
sign(KA_{n-1}) = (-1)^{n-1} n \sum_{d | n,\ n/d\ \text{odd}} d^3.
$$

We can again express the signatures of the KA_{n-1} in terms of modular forms (notations as in 2.3.15).

(3)
$$
\sum_{n=0}^{\infty} sign(KA_n)(-q)^n = \frac{d\epsilon}{dq}.
$$

Proof: As in 2.4.12 only the terms with $\nu = (n_1^{n/n_1})$, $\mu_{n/n_1} = (n/n_1)$ give a contribution to the χ_y-genus. So we get

$$
\chi_{-y}(KA_{n-1}) = \sum_{n_1 | n} n_1^4 \frac{(1 - x^{n/n_1})(1 - y^{n/n_1})}{(1-x)(1-y)} \Bigg|_{x=1}.
$$

(1) follows by easy computation and (2) by putting $y = -1$. (3) is obvious from the definition of ϵ. $\square$

By applying the same argument to the case of a geometrically ruled surface over an elliptic curve we get that $sign(KS_{n-1}) = 0$. This was however clear from the beginning as the dimension of KS_{n-1} is not divisible by 4. It seems remarkable that in all cases the signatures and the Euler numbers can be expressed in terms of the coefficients of the q-development of modular forms. For the first few of the $\chi_{-y}(KA_{n-1})$ we get:

$$\chi_{-y}(KA_1) = 2 + 20y + 2y^2,$$
$$\chi_{-y}(KA_2) = 3 + 6y + 90y^2 + 6y^3 + 3y^4,$$
$$\chi_{-y}(KA_3) = 4 + 8y + 44y^2 + 336y^3 + 44y^4 + 8y^5 + 4y^6,$$
$$\chi_{-y}(KA_4) = 5 + 10y + 15y^2 + 20y^3 + 650y^4 + 20y^5 + 15y^6 + 10y^7 + 5y^8,$$
$$\chi_{-y}(KA_5) = 6 + 12y + 18y^2 + 72y^3 + 288y^4 + 1800y^5 + 288y^6 + 72y^7 + 18y^8,$$
$$+ 12y^9 + 6y^{10}.$$

Let b_+ be the number of positive eigenvalues of the intersection form on the middle cohomology and b_- the number of negative ones. Then we get the following table:

n	$b_{2n}(KA_n)$	$sign(KA_n)$	$b_+(KA_n)$	$b_-(KA_n)$
1	22	-16	3	19
2	108	84	96	12
3	458	-256	101	357
4	1046	630	838	208
5	3748	-1320	1214	2534
6	7870	2408	5139	2731
7	25524	-4096	10714	14810
8	67049	6813	36931	30118
9	198270	-10080	94095	104175
10	538070	14652	276361	251709

We can also determine the Chern numbers of KA_2:

$$c_1^4 = 0, \quad c_1^2 c_2 = 0, \quad c_1 c_3 = 0,$$
$$c_4 = 108,$$
$$c_2^2 = 756,$$

This is true because $c_1 = -K_{KA_{n-1}} = 0$ and

$$sign(KA_{n-1}) = \frac{1}{45}(7p_2(KA_2) - p_1^2(KA_2)) = 84,$$

$$p_1(KA_2) = (c_1^2 - 2c_2)(KA_2) = -2c_2(KA_2),$$

$$p_2(KA_2) = (2c_4 - 2c_1c_3 + c_2^2)(KA_2) = 216 + c_2^2(KA_2).$$

2.5. The Betti numbers of varieties of triangles

Let X be a smooth projective variety of dimension d over a field k. For $d \geq 3$ and $n \geq 4$ the Hilbert scheme $X^{[n]}$ is singular. However $X^{[3]}$ is smooth for all $d \in \mathbb{N}$. In this section we want to compute the Betti numbers of $X^{[3]}$. $X^{[3]}$ can be viewed as a variety of unordered triangles on X. We also consider a number of other varieties of triangles on X, some of which have not yet appeared in the literature. As far as this is not yet known, we show that all these varieties are smooth. We study the relations between these varieties and compute their Betti numbers using the Weil conjectures.

Definition 2.5.1. [Elencwajg-Le Barz (5)] Let $\widetilde{\mathrm{Hilb}}^{n}(X) \subset X^{[n-1]} \times X^{[n]}$ be the reduced subvariety defined by

$$\widetilde{\mathrm{Hilb}}^{n}(X) = \left\{ (Z_{n-1}, Z_n) \in X^{[n-1]} \times X^{[n]} \mid Z_{n-1} \subset Z_n \right\}.$$

Here we will be interested in $\widetilde{\mathrm{Hilb}}^{3}(X)$. Let $i : \widetilde{\mathrm{Hilb}}^{3}(X) \longrightarrow X^{[2]} \times X^{[3]}$ be the embedding. If one interprets $X^{[3]}$ as a variety of unordered triangles on X, then $\widetilde{\mathrm{Hilb}}^{3}(X)$ parametrizes triangles Z_3 with a marked side Z_2. In the case $k = \mathbb{C}$ it was shown in [Elencwajg-Le Barz (5)] that $\widetilde{\mathrm{Hilb}}^{3}(X)$ is smooth. $\widetilde{\mathrm{Hilb}}^{3}(X)$ represents the contravariant functor from the category of $\underline{\mathrm{Schln}}_k$ locally noetherian k-schemes to the category $\underline{\mathrm{Ens}}$ of sets

$$\widetilde{\mathcal{Hilb}}^{3}(X) : \underline{\mathrm{Schln}}_k \longrightarrow \underline{\mathrm{Ens}};$$
$$T \longmapsto \left\{ (Z_2, Z_3) \in X^{[2]}(T) \times X^{[3]}(T) \mid Z_2 \subset Z_3 \right\},$$
$$(\phi : T_1 \longrightarrow T_2) \longmapsto ((Z_2, Z_3) \mapsto ((1_X \times \phi)^{-1}(Z_2), (1_X \times \phi)^{-1}(Z_3)).$$

So for a smooth variety X over $\mathbb{C}$ and a reduction X_0 of X modulo q the variety $\widetilde{\mathrm{Hilb}}^{3}(X_0)$ is a reduction of $\widetilde{\mathrm{Hilb}}^{3}(X)$ modulo q. Let

$$\bar{p}_2 : \widetilde{\mathrm{Hilb}}^{3}(X) \longrightarrow X^{[3]}$$

be the projection. For any partition ν of 3 (i.e. $\nu = (1,1,1)$, $J\nu = (2,1)$, $J\nu = (3)$) we put

$$\widetilde{\mathrm{Hilb}}^{3}_{\nu}(X) := \bar{p}_2^{-1}(X^{[3]}_{\nu}).$$

In [Elencwajg-Le Barz (5)] a residual point of a pair $(Z_2, Z_3) \in \widetilde{\mathrm{Hilb}}^{3}(X)$ is defined.

Definition 2.5.2. [Elencwajg-Le Barz (5)] Let

$$(Z_{n-1}, Z_n) \in X^{[n-1]} \times X^{[n]}.$$

Let I_{n-1} be the ideal of Z_{n-1} in $\mathcal{O}_{Z_n}$. Then the *residual point* $res(Z_{n-1}, Z_n) \in X$ is the point whose ideal in $\mathcal{O}_{Z_n}$ is the annihilator $Ann(I_{n-1}, \mathcal{O}_{Z_n})$ of I_{n-1} in $\mathcal{O}_{Z_n}$.

Elencwajg and Le Barz show that the map $(Z_{n-1}, Z_n) \mapsto res(Z_{n-1}, Z_n)$ gives a morphism $res : \widetilde{\mathrm{Hilb}}^n(X) \longrightarrow X$, if the ground field is $\mathbf{C}$. We show this for an arbitrary field.

Lemma 2.5.3. *The map* $(Z_{n-1}, Z_n) \mapsto res(Z_{n-1}, Z_n)$ *defines a morphism* $res :$ $\widetilde{\mathrm{Hilb}}^n(X) \longrightarrow X$.

Proof: Let T be an integral noetherian scheme and $(Z_{n-1}, Z_n) \in \widetilde{\mathcal{Hilb}}^n(X)(T)$. Let I be the ideal sheaf of Z_{n-1} in $\mathcal{O}_{Z_n}$. Then for all $t \in T$ the dimension of the annihilator $Ann(I_t, \mathcal{O}_{Z_n,t})$ is 1, so $Ann(I, \mathcal{O}_{Z_n})$ defines a subscheme $res(Z_{n-1}, Z_n) \subset Z_n$, which is flat of degree 1 over T, i.e a T-valued point of X. So res is given by a morphism of functors. $\square$

Remark 2.5.4. We can also describe the residual point as follows: for $(Z_{n-1}, Z_n) \in \widetilde{\mathrm{Hilb}}^3(X)$ the zero-cycle $\omega_n(Z_n) - \omega_{n-1}(Z_{n-1})$ is $[x]$ for some point $x \in X$ and $res(Z_{n-1}, Z_n) = x$. If we consider $\widetilde{\mathrm{Hilb}}^n(X)$ as a variety of triangles with a marked side, then res maps such a triangle to the vertex opposite to the marked side.

Via
$$i_1 := res \times i : \widetilde{\mathrm{Hilb}}^3(X) \longrightarrow X \times X^{[2]} \times X^{[3]}$$
we will in future consider $\widetilde{\mathrm{Hilb}}^3(X)$ as a subvariety of $X \times X^{[2]} \times X^{[3]}$:

$$\widetilde{\mathrm{Hilb}}^3(X) = \left\{ (x, Z_2, Z) \in X \times X^{[2]} \times X^{[3]} \mid x \subset Z, \ Z_2 \subset Z, \ res(Z_2, Z) = x \right\}$$

This means we consider $\widetilde{\mathrm{Hilb}}^3(X)$ as a variety of triangles with a side and the opposing vertex marked. Let

$$\bar{p}_1 : \widetilde{\mathrm{Hilb}}^3(X) \longrightarrow X,$$
$$\bar{p}_2 : \widetilde{\mathrm{Hilb}}^3(X) \longrightarrow X^{[2]},$$
$$\bar{p}_3 : \widetilde{\mathrm{Hilb}}^3(X) \longrightarrow X^{[3]},$$
$$\bar{p}_{1,2} : \widetilde{\mathrm{Hilb}}^3(X) \longrightarrow X \times X^{[2]},$$
$$\bar{p}_{1,3} : \widetilde{\mathrm{Hilb}}^3(X) \longrightarrow X \times X^{[3]}$$

be the projections. From the definitions we can see that the support of the image of $\bar{p}_{1,3}$ coincides with the support of the universal subscheme $Z_3(X)$. As $\widetilde{\mathrm{Hilb}}^3(X)$ is reduced, this defines a morphism

$$\bar{p}_{1,3} : \widetilde{\mathrm{Hilb}}^3(X) \longrightarrow Z_3(X).$$

This morphism is birational, as its restriction gives an isomorphism from $(\widetilde{\mathrm{Hilb}}^3(X))_{(1,1,1)}$ to a dense open subset of $Z_3(X)$. So $\bar{p}_{1,3} : \widetilde{\mathrm{Hilb}}^3(X) \longrightarrow Z_3(X)$ is a canonical resolution of $Z_3(X)$. We can consider $Z_3(X)$ as the variety of triangles with a marked vertex. Then $\bar{p}_{1,3}$ is given by forgetting the marked side.

$$\bar{p}_{1,2} : \widetilde{\mathrm{Hilb}}^3(X) \longrightarrow X \times X^{[2]}$$

is birational, as it gives an isomorphism of the dense open subvariety $(\widetilde{\mathrm{Hilb}}^3(X))_{(1,1,1)}$ onto ints image. Let $Z_2(X) \subset X \times X^{[2]}$ be the universal sub-scheme. As a set $Z_2(X)$ is given by

$$Z_2(X) = \left\{ (x, Z) \subset X \times X^{[2]} \,\middle|\, x \subset Z \right\}.$$

One can also verify easily that it carries the reduced induced structure and that it can be described as $X \times X$ blown up allong the diagonal. Let

$$w : X \times X^{[2]} \longrightarrow Z_3(X)$$

be the rational map which is defined on the open dense subvariety $(X \times X^{[2]}) \setminus Z_2(X)$ by $w\big((x, Z)\big) := (x, x \cup Z)$. Then obviously the diagram

$$\widetilde{\mathrm{Hilb}}^3(X)$$

$$\Big\downarrow{\scriptstyle \bar{p}_{1,2}} \qquad \searrow{\scriptstyle \bar{p}_{1,3}}$$

$$X \times X^{[2]} \quad \dashrightarrow^{\,w\,} \quad Z_3(X)$$

commutes. So $\bar{p}_{1,3} : \widetilde{\mathrm{Hilb}}^3(X) \longrightarrow Z_3(X)$ is a natural resolution of the indeterminacy of w. We will see later that $\widetilde{\mathrm{Hilb}}^3(X)$ is the blow up of $X \times X^{[2]}$ along $Z_2(X)$.

The varieties of complete triangles on X.

Semple [Semple (1)] has constructed a variety of complete triangles on $\mathbf{P}_2$. This variety has been studied and its Chow ring was determined in [Roberts (1)], [Roberts-Speiser (1),(2),(3),(4)], [Collino-Fulton (1)]. (The Chow ring coincides with the cohomology ring in case $k = \mathbf{C}$). Le Barz has generalized this construction in [Le Barz (10)] to general projective varieties and shown that the resulting varieties of complete triangles are smooth. Keel [Keel (1)] also gave a functorial construction of these varieties. Let X be a smooth projective variety of dimension d over a field k. We want to define other varieties of complete triangles. Because of this we call the variety defined by Le Barz the variety of complete ordered triangles on X. We also want to show that our varieties of complete triangles are smooth by using results from [Le Barz (10)].

Definition 2.5.5. [Le Barz (10)] Let X be a smooth projective variety over a field k. The variety $\widehat{H}^3(X)$ of complete ordered triangles on X is the closed subvariety of $X^3 \times (X^{[2]})^3 \times X^{[3]}$ defined by

$$\widehat{H}^3(X) = \left\{ (x_1, x_2, x_3, Z_1, Z_2, Z_3, Z) \in (X^3 \times (X^{[2]})^3 \times X^{[3]}) \;\middle|\; \begin{array}{c} x_i, x_j \subset Z_l; \; Z_i \subset Z; \\ x_l = res(x_i, Z_j) = res(Z_l, Z) \\ \text{for all permutations} \\ (i, j, l) \text{ of } (1, 2, 3) \end{array} \right\}.$$

In [Le Barz (10)] $\widehat{H}^3(X)$ is shown to be smooth for X a smooth variety over $\mathbf{C}$. $\widehat{H}^3(X)$ represents the obvious functor $\widehat{\mathcal{H}}^3(X) : \underline{\mathrm{Schln}}_k \longrightarrow \underline{\mathrm{Ens}}$:

$$\widehat{\mathcal{H}}^3(X)(T) = \left\{ (x_1, x_2, x_3, Z_1, Z_2, Z_3, Z) \in (X^3 \times (X^{[2]})^3 \times X^{[3]})(T) \;\middle|\; \begin{array}{c} x_i, x_j \subset Z_l; \; Z_i \subset Z; \\ x_l = res(x_i, Z_j) = res(Z_l, Z) \\ \text{for all permutations} \\ (i, j, l) \text{ of } (1, 2, 3) \end{array} \right\}$$

(see also [Collino-Fulton (1) rem. (5)]). So if X is a smooth projective variety over $\mathbf{C}$ and X_0 is a good reduction of X modulo q, then $\widehat{H}^3(X_0)$ is a reduction of $\widehat{H}^3(X)$ modulo q. Let

$$j : \widehat{H}^3(X) \longrightarrow X^3 \times (X^{[2]})^3 \times X^{[3]}$$

be the embedding. Let

$$\hat{p}_1 : \widehat{H}^3(X) \longrightarrow X^3,$$
$$\hat{p}_2 : \widehat{H}^3(X) \longrightarrow (X^{[2]})^3,$$
$$\hat{p}_3 : \widehat{H}^3(X) \longrightarrow X^{[3]},$$

be the projections. From the stratification of $X^{[3]}$ we get one of $\widehat{H}^3(X)$. Let ν be a partition of 3. Then we put

$$(\widehat{H}^3(X))_\nu := \hat{p}_3^{-1}(X_\nu^{[3]}).$$

We can view the x_i as the vertices of the triangle Z and Z_i as the side opposite to x_i. Thus $\widehat{H}^3(X)$ parametrizes the complete ordered triangles on X (i.e. together with a triangle we are given all its vertices and all its sides together with an ordering). The projection $\hat{p}_1 : \widehat{H}^3(X) \longrightarrow X^3$ is birational.

Definition 2.5.6. [Le Barz (10)] For a pair (i,j) satisfying $1 \le i < j \le 3$ let

$$\Delta_{i,j} := \left\{ (Z_1, Z_2, Z_3) \in (X^{[2]})^3 \,\Big|\, Z_i = Z_j \right\} \subset (X^{[2]})^3$$

be the diagonal between the ith and j^{th} factors. Let

$$\delta_1 := \left\{ (x_1, x_2, x_3) \in X^3 \,\Big|\, x_1 = x_2 = x_3 \right\}$$

be the small diagonal in X^3, and

$$\delta_2 := \left\{ (Z_1, Z_2, Z_3) \in (X^{[2]})^3 \,\Big|\, Z_1 = Z_2 = Z_3 \right\}$$

the small diagonal in $(X^{[2]})^3$. Then we put

$$E_{i,j}(X) := \hat{p}_2^{-1}(\Delta_{i,j}),$$
$$D_1^2(X) := (\hat{p}_1 \times \hat{p}_2)^{-1}(\delta_1 \times \delta_2).$$

In [Le Barz (10)] these varieties are shown to be smooth for X a smooth variety over **C**. The $E_{i,j}(X)$ are irreducible divisors in $\widehat{H}^3(X)$. $D_1^2(X)$ is the variety of second order data on X, which we want to study in more detail in chapter 3.

For $x \in X$ let $\mathbf{m}_{X,x}$ be the maximal ideal in the local ring $\mathcal{O}_{X,x}$ and

$$q_x : \mathbf{m}_{X,x} \longrightarrow \mathbf{m}_{X,x}/\mathbf{m}_{X,x}^2$$

the natural projection. We can describe the subscheme $Z_{(1,2)}(X) \subset X^{[3]}$ (cf. section 2.1) as the closed reduced subvariety given by

$$\left\{ Z \in X^{[3]} \,\middle|\, \begin{array}{c} supp(Z) = x \text{ for an } x \in X, \text{ and there is a} \\ \text{2-codimensional linear subspace } V \subset \mathbf{m}_{X,x}/\mathbf{m}_{X,x}^2 \text{ such that} \\ \text{the ideal } I_Z \text{ of } Z \text{ in } \mathcal{O}_{X,x} \text{ is of the form } I_Z = q_x^{-1}(V) \end{array} \right\}.$$

Obviously $Z_{(1,2)}(X)$ is isomorphic to the Grassmannian bundle $Grass(2, T_X^*)$ of two-dimensional quotients of the cotangent bundle of X. We put

$$E := \hat{p}_3^{-1}(Z_{(1,2)}(X))$$
$$= \left\{ (x,x,x,Z_1,Z_2,Z_3,Z) \,\middle|\, \begin{array}{c} x \in X; \ Z_1, Z_2, Z_3 \in X^{[2]}; \ Z \in Z_{(1,2)}(X); \\ supp(Z) = x; Z_1, Z_2, Z_3 \subset Z \end{array} \right\}.$$

Let $Z \in Z_{(1,2)}(X)$, $x := supp(Z)$. Then the ideal I_Z of Z in $\mathcal{O}_{X,x}$ is of the form $I_Z = q_x^{-1}(V)$ for a suitable 2-codimensional linear subspace V of $\mathbf{m}_{X,x}/\mathbf{m}_{X,x}^2$. Let

$$q_Z : \mathbf{m}_{X,x} \longrightarrow \mathbf{m}_{X,x}/I_Z$$

be the natural projection. The ideals I_{Z_2} of subschemes Z_2 of length 2 of Z are given exactly by the $q_Z^{-1}(W)$ for the one-dimensional linear subspaces W of $\mathbf{m}_{X,x}/I_Z$. Let

$$\pi : Z_{(1,2)}(X) = Grass(2, T_X^*) \longrightarrow X$$

be the projection. Then the subschemes $Z_2 \subset Z$ of length 2 are given by the one-dimensional linear subspaces of the fibre of the tautological subbundle T_1 of $\pi^*(T_X^*)$ over the point V. Thus we get

Remark 2.5.7. $E \cong \mathbf{P}(T_1) \times_{Grass(2, T_X^*)} \mathbf{P}(T_1) \times_{Grass(2, T_X^*)} \mathbf{P}(T_1)$.

Proposition 2.5.8. *Let X be a smooth projective variety over* **C**. *Then* $\bar{p}_{1,2} : \widetilde{\mathrm{Hilb}}^3(X) \longrightarrow X \times X^{[2]}$ *is the blow up along $Z_2(X)$.*

Proof:

$\bar{p}_{1,2} : \widetilde{\mathrm{Hilb}}^3(X) \longrightarrow X \times X^{[2]}$ is an isomorphism over $(X \times X^{[2]}) \setminus Z_2(X)$. Let $F := \bar{p}_{1,2}^{-1}(Z_2(X))$. Then F can be described as the set:

$$F = \left\{ (x, Z_1, Z) \in X \times X^{[2]} \times X^{[3]} \mid x_1 \subset Z_1, \ Z_1 \subset Z, \ res(Z_1, Z) = x_1 \right\}.$$

Let

$$p_{1,4,7} : \widehat{H}^3(X) \longrightarrow X \times X^{[2]} \times X^{[3]}$$
$$(x_1, x_2, x_3, Z_1, Z_2, Z_3, Z) \longmapsto (x_1, Z_1, Z)$$

be the projection. We see immediately that the image of this morphism is $\widetilde{\mathrm{Hilb}}^3(X)$ so we get a morphism

$$\check{p}_{1,4,7} : \widehat{H}^3(X) \longrightarrow \widetilde{\mathrm{Hilb}}^3(X).$$

Let

$$(x_1, x_2, x_3, Z_1, Z_2, Z_3, Z) \in E_{1,2}(X).$$

Then we have $Z_1 = Z_2$ and thus $x_1 = x_2$. So we get $x_1 \subset Z_1$. We see that

$$\check{p}_{1,4,7}(E_{1,2}(X)) \subset F.$$

So we get a morphism $q : E_{1,2}(X) \longrightarrow F$. Let $(x_1, Z_1, Z) \in F$. We put $x_2 := x_1$, $x_3 := res(x_1, Z_1)$. If $supp(Z)$ consists of two points, we see that $x_1 \neq x_3$ and $Z = Z_3 \cup x_3$ for a unique subscheme Z_3 of length 2 with support x_1. If $supp(Z)$ is a point but Z does not lie in $Z_{(1,2)}(X)$, then Z has a unique subscheme Z_3 of length 2. In both cases we get

$$q^{-1}(x_1, Z_1, Z) = \{(x_1, x_2, x_3, Z_1, Z_1, Z_3, Z)\}.$$

If Z lies in $Z_{(1,2)}(X)$, then it is given by a two-dimensional quotient W of the cotangent space $T_X^*(x_1)$ of X at x_1, and the subschemes Z_3 of Z are given by the one-dimensional quotients V of W. So we get

$$q^{-1}(x_1, Z_1, Z) = \left\{(x_1, x_2, x_3, Z_1, Z_1, Z_3, Z) \,\middle|\, Z_3 \subset Z\right\} \cong \mathbf{P}_1.$$

Putting things together we see that q is onto and a bijection over the open set $F \backslash \bar{p}_3^{-1}(Z_{(1,2)}(X))$. As $E_{1,2}(X)$ is an irreducible divisor of $\hat{H}^3(X)$, F is an irreducible divisor on $\widetilde{\mathrm{Hilb}}^3(X)$. Let

$$\varepsilon : X \widetilde{\times} X^{[2]} \longrightarrow X \times X^{[2]}$$

be the blow up of $X \times X^{[2]}$ along $Z_2(X)$. Let $\mathcal{I}$ be the ideal of $Z_2(X)$ in $X \times X^{[2]}$. From $\bar{p}_{1,2}^{-1}(Z_2(X)) = F$ we get that $\bar{p}_{1,2}^{-1}\mathcal{I} \cdot \mathcal{O}_{\widetilde{\mathrm{Hilb}}^3(X)}$ is the invertible sheaf corresponding to F. By the universal property of the blow up (cf. e.g. [Hartshorne (2), II. prop.7.14]) there is a morphism

$$g : \widetilde{\mathrm{Hilb}}^3(X) \longrightarrow X \widetilde{\times} X^{[2]}$$

such that the diagram

$$
\begin{array}{ccc}
\widetilde{\mathrm{Hilb}}^3(X) & \xrightarrow{\quad g \quad} & X \widetilde{\times} X^{[2]} \\[2ex]
& \searrow{\scriptstyle \bar{p}_{1,2}} & \downarrow{\scriptstyle \varepsilon} \\[2ex]
& & X \times X^{[2]}
\end{array}
$$

commutes. g is a birational morphism. By [Hartshorne (1) II Thm. 7.17] g is the blow up of a subscheme of $X \times X^{[2]}$. g is an isomorphism outside F, F is irreducible, and the image $g(F)$ is the exceptional divisor of the blow up $\varepsilon : X \times X^{[2]} \longrightarrow X \times X^{[2]}$. Thus g is an isomorphism and the result follows.   □

In a joint work with Barbara Fantechi [Fantechi-Göttsche (1)] we use proposition 2.5.8 to compute the ring structure cohomology ring $H^*(X^{[3]}, \mathbf{Q})$ of the Hilbert scheme of three points on a smooth projective variety X of arbitrary dimension in terms of the cohomology ring of X. We also compute the cohomology ring of $\widetilde{\mathrm{Hilb}}^3(X)$.

Proposition 2.5.8 also follows from [Kleiman (3)] thm 2.8. I have learned that Ellingsrud [Ellingsrud (1)] has proven independently the following: if S is a smooth surface, the blow up of $S \times S^{[n]}$ along the universal family

$$Z_n(S) = \left\{ (x, Z) \in S \times S^{[n]} \ \middle| \ x \in Z \right\}$$

is a smooth variety mapping surjectively to $S^{[n+1]}$ (proposition 2.5.8 is essentially the case $n = 2$ of this).

One can see easily that $E_{1,2}(X)$ is obtained from F by blowing up along $\bar{p}_3^{-1}(Z_{(1,2)}(X))$.

Definition 2.5.9. For all $n \in I\!N$ let

$$\Phi_{X,n} : X^n \longrightarrow X^{(n)},$$
$$\Phi_{X^{[2]},n} : (X^{[2]})^n \longrightarrow (X^{[2]})^{(n)}$$

be the quotient morphisms. Then let

$$\widehat{X}^{[3]} \subset X^{(3)} \times (X^{[2]})^{(3)} \times X^{[3]}$$

be the image of $\widehat{H}^3(X)$ under

$$\Phi_{X,3} \times \Phi_{X^{[2]},3} \times \Phi_{X^{[3]}} : X^3 \times (X^{[2]})^3 \times X^{[3]} \longrightarrow X^{(3)} \times (X^{[2]})^{(3)} \times X^{[3]}$$

with the reduced induced structure. Let $\pi_1 : \widehat{H}^3(X) \longrightarrow \widehat{X}^{[3]}$ be the restriction of this morphism to $\widehat{H}^3(X) \subset X^3 \times (X^{[2]})^3 \times X^{[3]}$.

The symmetric group $G(3)$ acts on $X^3 \times (X^{[2]})^3 \times X^{[3]}$ by permuting the factors in X^3 and $(X^{[2]})^3$ simultaniously. $\pi_1 : \widehat{H}^3(X) \longrightarrow \widehat{X}^{[3]}$ is the quotient morphism with respect to the induced action on $\widehat{H}^3(X)$. We can consider $\widehat{X}^{[3]}$ as a variety of complete unordered triangles on X, as together with a triangle $Z \in X^{[3]}$ we are given all its vertices $[x_1] + [x_2] + [x_3]$ and all the sides $[Z_1] + [Z_2] + [Z_3]$ (however without an ordering). The projection

$$p_3 : X^{(3)} \times (X^{(2)})^{(3)} \times X^{[3]} \longrightarrow X^{[3]}$$

induces a birational morphism

$$p : \widehat{X}^{[3]} \longrightarrow X^{[3]}$$

(p is an isomorphism over the open dense subset $X^{[3]}_{(1,1,1)}$). We can again give a stratification of $\widehat{X}^{[3]}$ by putting

$$\widehat{X}^{[3]}_\nu := p^{-1}(X^{[3]}_\nu)$$

for all partitions ν of 3. We put

$$\bar{E} := p^{-1}(Z_{(1,2)}(X))$$
$$= \left\{ (3[x], [Z_1] + [Z_2] + [Z_3], Z) \left| \begin{array}{c} x \in X; \ Z_1, Z_2, Z_3 \in X^{[2]}; \ Z \in Z_{(1,2)}(X); \\ supp(Z) = x; Z_1, Z_2, Z_3 \subset Z \end{array} \right. \right\}.$$

Then we have $\bar{E} = \pi_1(E)$. The action of $G(3)$ on $\widehat{H}^3(X)$ maps E to itself. The induced operation of $G(3)$ on E is by permuting Z_1, Z_2, Z_3 and $\bar{E}$ is the quotient. So we get from remark 2.5.7:

Remark 2.5.10.

$$\bar{E} \cong (\mathbf{P}(T_1) \times_{Grass(2,T_X^*)} \mathbf{P}(T_1) \times_{Grass(2,T_X^*)} \mathbf{P}(T_1))/G(3)$$
$$= \mathbf{P}(Sym^3(T_1)).$$

Proposition 2.5.11. *Let X be a smooth projective variety over $\mathbf{C}$. Then*

(1) $\widehat{X}^{[3]}$ *is smooth.*

(2) $p: \widehat{X}^{[3]} \longrightarrow X^{[3]}$ *is the blow up along* $Z_{(1,2)}(X)$.

Proof: It is clear that p is an isomorphism over the open dense subset $(X^{[3]})_{(1,1,1)}$. Let $Z = (Z_2 \cup x) \in X^{[3]}_{(2,1)}$, i.e.

$$Z_2 \in X^{[2]}_{(2)}, \ x \in X, \ y := supp(Z_2) \neq x.$$

Then we have

$$p^{-1}(Z) = \left\{ (2[y] + [x], 2[(x \cup y)] + [Z_2], Z) \right\}.$$

Now let

$$Z \in (X^{[3]}_{(3)} \setminus Z_{(1,2)}(X)) = Z_{(1,1,1)}(X)$$

and $x := supp(Z)$. Then the ideal of Z in $\mathcal{O}_{X,x}$ is given by

$$I_Z = (x_1^3, x_2, \ldots, x_d)$$

for suitable local parameters $x_1, x_2, \ldots, x_d$. The subscheme Z_2 given by

$$I_{Z_2} = (x_1^2, x_2, \ldots, x_d)$$

is the only subscheme of length 2 in Z, and we have

$$p^{-1}(Z) = \left\{ (3[x], 3[Z_2], Z) \right\}.$$

As $X^{[3]}$ is smooth, p is an isomorphism over $X^{[3]} \setminus Z_{(1,2)}(X)$ by Zariski's main theorem [Hartshorne (2), V. 5.2]. Now we show (1). As p is an isomorphism over $X^{[3]} \setminus Z_{(1,2)}(X)$, it is enough to prove the smoothness at the points of $\bar{E} = \pi_1(E)$. Le Barz has given analytic local coordinates around any point $e \in E$ and so proved the smoothness of $\widehat{H}^3(X)$. To simplify notations we will assume that the dimension of X is 3. The argument for general dimension d is completely analogous, only more difficult to write down. Now let

$$e = (o, o, o, Z_1, Z_2, Z_3, Z) \in E$$

and $\bar{e} := \pi_1(E)$. We choose local coordinates x, y, z on X centered at o. By choosing x, y, z suitably we can assume that

$$I_Z := (x^2, xy, y^2, z)$$

is the ideal of Z and that

$$I_{Z_1} := (x^2, y, z)$$

is the ideal of Z_1. We have to distinguish 3 cases:

(a) $Z_1 = Z_2 = Z_3$. Then Le Barz constructs the chart

$$(r_1, s_1, t_1, c_1, c_2, c_3, v, \rho, \sigma)$$

around e as follows: let

$$e' := (o_1, o_2, o_3, Z_1', Z_2', Z_3', Z')$$

be a point of $\widehat{H}^3(X)$ near e. The ideal $I_{Z'}$ of Z' can be written as:

$$I_{Z'} = (x^2 + ux + vy + w, xy + u'x + v'y + w', y^2 + u''x + v''y + w'', z + \rho x + \sigma y + \theta)$$

for suitable $u, v, w, u', v', u'', v'', w'', \rho, \sigma, \theta$. Let

$$(r_1, s_1, t_1), \ (r_2, s_2, t_2), \ (r_3, s_3, t_3)$$

be the coordinates of the points o_1, o_2, o_3. The ideal $I_{Z_i'}$ of Z_i' can be written

$$I_{Z_i'} = (x^2 + a_i x + b_i, -y + c_i x + d_i, -z + e_i x + f_i)$$

for suitable $a_i, b_i, c_i, d_i, e_i, f_i$. Now Le Barz shows that all other constants can be computed from $r_1, s_1, t_1, c_1, c_2, c_3, v, \rho, \sigma$, and that $(r_1, s_1, t_1, c_1, c_2, c_3, v, \rho, \sigma)$ is a local chart of $\widehat{H}^3(X)$ around e. Because of the symmetry we can replace r_1, s_1, t_1 by r_2, s_2, t_2 or r_3, s_3, t_3 and so also by

$$r := r_1 + r_2 + r_3, \ s := s_1 + s_2 + s_3, \ t := t_1 + t_2 + t_3.$$

So we get the local chart $(r, s, t, c_1, c_2, c_3, v, \rho, \sigma)$ around e. With respect to this chart the action of $\tau \in G(3)$ on $\hat{H}^3(X)$ is given by

$$\tau(r) = r, \ \tau(s) = s, \ \tau(t) = t,$$
$$\tau(c_i) = c_{\tau(i)},$$
$$\tau(v) = v, \ \tau(\rho) = \rho, \ \tau(\sigma) = \sigma.$$

So we see that $(r, s, t, c_1 + c_2 + c_3, c_1 c_2 + c_1 c_3 + c_2 c_3, c_1 c_2 c_3, v, \rho, \sigma)$ are local coordinates of $\hat{X}^{[3]}$ around $\bar{e}$.

(b) $Z_1 = Z_2 \neq Z_3$. In this case we can choose the local coordinates x, y, z in such a way that the ideal I_{Z_3} of Z_3 is given by

$$I_{Z_3} = (x^2, y - x, z).$$

So the ideal $I_{Z_3'}$ of Z_3' is of the form

$$I_{Z_3'} = (x^2 + ax + b, -y + (\gamma + 1)x + d, -z + ex + f).$$

By [Le Barz 10] $(r, s, t, c_1, c_2, \gamma, v, \rho, \sigma)$ are local coordinates around e. The stabilizer of the operation around e is $G(3)_e = \{1, (1, 2)\}$. We can choose the coordinate neighbourhood so small that we have

$$\tau(U) \cap U \neq \emptyset \iff \tau \in G(3)_e.$$

$r, s, t, \gamma, v, \rho, \sigma$ are fixed by the action of $G(3)_e$, and we have

$$(1, 2)(c_1) = c_2, \ (1, 2)(c_2) = c_1.$$

So $(r, s, t, c_1 + c_2, c_1 c_2, \gamma, v, \rho, \sigma)$ form a local chart of $\hat{X}^{[3]}$ at $\bar{e}$.

(c) The Z_i are pairwise distict. We can assume that the ideals I_{Z_2}, I_{Z_3} of Z_2, Z_3 are of the form

$$I_{Z_2} = (x^2, -y + x, z),$$
$$I_{Z_3} = (x^2, x + y, z).$$

Then the ideals $I_{Z_2'}, I_{Z_3'}$ of Z_2', Z_3' can be written in the form

$$I_{Z_2'} = (x^2 + a_2 x + b_2, -y + (\gamma + 1)x + d_2, -z + e_2 x + f_2),$$
$$I_{Z_3'} = (x^2 + a_3 x + b_3, -y + (\gamma' - 1)x + d_3, -z + e_3 x + f_3).$$

Le Barz shows that $(r, s, t, c_1, \gamma, \gamma', v, \rho, \sigma)$ form a local chart of $\hat{H}^3(X)$ around e. The stabilizer of the action of $G(3)$ at e is $G(3)_e = \{1\}$. Again we can choose the coordinate neighbourhood around e to be so small that we have

$$\tau(U) \cap U \neq \emptyset \iff \tau = 1.$$

Then $(r, s, t, c_1, \gamma, \gamma', v, \rho, \sigma)$ is also a local chart of $\widehat{X}^{[3]}$ at $\bar{e}$. Putting things together we have proved (1).

We already know from remark 2.5.10 that $\bar{E} := p^{-1}(Z_{(1,2)}(X))$ is a locally trivial $\mathbf{P}_3$-bundle over $Z_{(1,2)}(X) = Grass(2, T_X^*)$. In particular $\bar{E}$ is an irreducible divisor on $\widehat{X}^{[3]}$. So we can complete the proof of (2) in the same way as that of proposition 2.5.8.　□

Keel [Keel (1)] has proved by a different method that the symmetric group $G(3)$ acts on $\widehat{H}^3(X)$ and that the quotient is the blowup of $X^{[3]}$ along $Z_{(1,2)}(X)$.

Let

$$\widehat{\mathrm{Hilb}}^3(X) \subset X \times X^{(2)} \times X^{[2]} \times (X^{[2]})^{(2)} \times X^{[3]}$$

be the scheme-theoretic image of $\widehat{H}^3(X)$ under $1_X \times \Phi_{X,2} \times 1_{X^{[2]}} \times \Phi_{X^{[2]},2} \times 1_{X^{[3]}}$ and let

$$\pi_2 : \widehat{H}^3(X) \longrightarrow \widehat{\mathrm{Hilb}}^3(X)$$

be the restriction of this morphism. $\mathbb{Z}/2\mathbb{Z}$ acts on $X^3 \times (X^{[2]})^3 \times X^{[3]}$ by permuting the last two factors in X^3 and $(X^{[2]})^3$ simultaniously. This action restricts to an action on $\widehat{H}^3(X)$. Let $\pi_2 : \widehat{H}^3(X) \longrightarrow \widehat{\mathrm{Hilb}}^3(X)$ be the quotient morphism. π_1 factorizes into

$$
\begin{array}{ccc}
\widehat{H}^3(X) & \xrightarrow{\ \pi_1\ } & \widehat{X}^{[3]} \\[2mm]
\Big\downarrow{\scriptstyle \pi_2} & \nearrow{\scriptstyle \pi_3} & \\[2mm]
\widehat{\mathrm{Hilb}}^3(X). & &
\end{array}
$$

We can view $\widehat{\mathrm{Hilb}}^3(X)$ as the variety of complete triangles on X with a marked vertex (or equivalently with a marked side). The projection

$$p_{1,3,5} : X \times X^{(2)} \times X^{[2]} \times (X^{[2]})^{(2)} \times X^{[3]} \longrightarrow X \times X^{[2]} \times X^{[3]}$$

restricts to a birational morphism

$$\check{p}_{1,3,5} : \widehat{\mathrm{Hilb}}^3(X) \longrightarrow \widetilde{\mathrm{Hilb}}^3(X).$$

Let $\bar{p}_3 : \widetilde{\mathrm{Hilb}}^3(X) \longrightarrow X^{[3]}$ be the projection. We put:

$$B(X) := \bar{p}_3^{-1}(Z_{(1,2)}(X)) \subset \widetilde{\mathrm{Hilb}}^3(X)$$

with the reduced induced structure. $B(X)$ is a $\mathbf{P}_1$-bundle over $Z_{(1,2)}(X) \cong Grass(2, T_X^*)$. In fact we can see in the same way as above that $B(X) = \mathbf{P}(T_1)$ holds, where T_1 is the tautological bundle on $Grass(2, T_X^*)$. We put

$$
\begin{aligned}
\tilde{E} &:= \check{p}_{1,3,5}^{-1}(B(X)) \\
&= \left\{ (x, 2[x], Z_1, [Z_2] + [Z_3], Z) \left| \begin{array}{l} x \in X;\ Z_1, Z_2, Z_3 \in X^{[2]};\ Z \in Z_{(1,2)}(X); \\ supp(Z) = x;\ Z_1, Z_2, Z_3 \subset Z \end{array} \right. \right\}.
\end{aligned}
$$

Then we have $\tilde{E} = \pi_2(E)$. The action of $\mathbb{Z}/2\mathbb{Z}$ on $\widehat{H}^3(X)$ restricts to an action on E by permuting Z_2, Z_3, and the quotient is $\tilde{E}$. So we get from remark 2.5.7:

Remark 2.5.12.

$$\tilde{E} \cong (\mathbf{P}(T_1) \times_{Grass(2,T_X^*)} \mathbf{P}(T_1) \times_{Grass(2,T_X^*)} \mathbf{P}(T_1))/(\mathbb{Z}/2\mathbb{Z})$$
$$= \mathbf{P}(T_1) \times_{Grass(2,T_X^*)} \mathbf{P}(\mathrm{Sym}^2(T_1)),$$

and the restriction $\check{p}_{1,3,5} : \tilde{E} \longrightarrow B(X)$ is the projection onto the first factor.

Proposition 2.5.13. *Let X be a smooth projective variety over* **C**. *Then*

(1) $\widetilde{\mathrm{Hilb}}^3(X)$ *is smooth.*

(2) $\check{p}_{1,3,5} : \widetilde{\mathrm{Hilb}}^3(X) \longrightarrow \widetilde{\mathrm{Hilb}}^3(X)$ *is the blow up along $B(X)$.*

Proof: $\check{p}_{1,3,5}$ is obviously an isomorphism over $\widetilde{\mathrm{Hilb}}^3(X)_{(1,1,1)}$. Let $(x, Z_2, Z) \in \widetilde{\mathrm{Hilb}}^3(X)_{(2,1)}$. Then there are two cases:

(α) $Z_2 = x \cup y$ for a point $y \neq x$ and $Z = W_2 \cup y$ for a subscheme W_2 of length 2 with $supp(W_2) = x$. Then we have

$$\check{p}_{1,3,5}^{-1}((x, Z_2, Z)) = \left\{ (x, [x] + [y], x \cup y, [x \cup y] + [W_2], W_2 \cup y) \right\}.$$

(β) $supp(Z_2) = y \neq x$. Then we have

$$\check{p}_{1,3,5}^{-1}((x, Z_2, Z)) = \left\{ (x, 2[y], Z_2, 2[x \cup y], x \cup Z_2) \right\}.$$

Now let

$$(x, Z_2, Z) \in \widetilde{\mathrm{Hilb}}^3(X)_{(3)} \setminus B(X).$$

Then Z_2 is the only subscheme of length 2 contained in Z. So we have

$$\check{p}_{1,3,5}^{-1}((x, Z_2, Z)) = \left\{ (x, 2[x], Z_2, 2[Z_2], Z) \right\}.$$

As $\widetilde{\mathrm{Hilb}}^3(X)$ is smooth, this shows that $\check{p}_{1,3,5}$ is an isomorphism over $\widetilde{\mathrm{Hilb}}^3(X) \setminus B(X)$. We now show (1). As above we only have to show the smoothness of $\widetilde{\mathrm{Hilb}}^3(X)$ in points of $\tilde{E}$. We again use the local charts of Le Barz around a point $e = (o, o, o, Z_1, Z_2, Z_3, Z) \in E$. Let $\tilde{e} := \pi_2(e)$. We use the same notations as in proposition 2.5.11. There are four cases:

(a) $Z_1 = Z_2 = Z_3$. We see anologously to the proof of proposition 2.5.11 that $(r, s, t, c_1, c_2 + c_3, c_2 c_3, v, \rho, \sigma)$ form a local chart of $\widetilde{\mathrm{Hilb}}^3(X)$ around $\tilde{e}$.

(b) $Z_1 \neq Z_2 = Z_3$. We switch the role of Z_1 and Z_3 in the case (b) in the proof of proposition 2.5.11. So we see immediately that $(r, s, t, \gamma, c_2 + c_3, c_2 c_3, v, \rho, \sigma)$ form a local chart around $\tilde{e}$.

(c) $Z_1 = Z_2 \neq Z_3$. We switch the role of Z_1 and Z_2 in (b) in 2.5.11. This way we see that $(r, s, t, c_1, \gamma, c_3, v, \rho, \sigma)$ form a local chart at $\tilde{e}$.

(d) Z_1, Z_2, Z_3 are pairwise distinct. Then $(r, s, t, c_1, \gamma, \gamma', v, \rho, \sigma)$ form a local chart near $\tilde{e}$.

We have proved (1). By remark 2.5.12 $\tilde{E} \longrightarrow B(X)$ is a locally trivial $\mathbf{P}_2$-bundle. In particular $\tilde{E}$ is an irreducible divisor on $\widehat{\mathrm{Hilb}}^3(X)$. Now (2) follows in the same way as in the proof of 2.5.8 and 2.5.11(2).   □

If we put our results together, we get the following diagram for the triangle varieties of a smooth projective variety X.

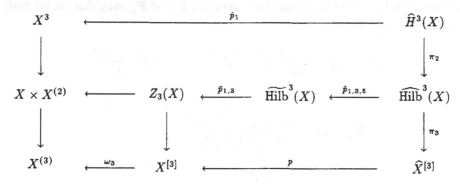

Here the horizontal arrows are birational morphisms.

Computation of the Betti numbers of some of these varieties

To compute the Betti numbers of some of these varieties we will again use the Weil conjectures. So we have to count their points over finite fields. First we look at the local situation. Let k be a field and $R = k[[x_1, \ldots, x_d]]$. As above $\mathrm{Hilb}^n(R)$ parametrizes the ideals of colength n in R.

Definition 2.5.14. For all $l \in I\!\!N$ let $W_3^l \subset (\mathrm{Hilb}^2(R))^l \times \mathrm{Hilb}^3(R)$ be the reduced closed subscheme defined by

$$W_3^l = \left\{ (I_1, \ldots, I_l, J) \in (\mathrm{Hilb}^2(R))^l \times \mathrm{Hilb}^3(R) \,\middle|\, I_1, \ldots, I_l \supset J \right\}.$$

Now let k be a finite field $I\!\!F_{\tilde{q}}$.

Lemma 2.5.15. *There is a finite field extension $I\!\!F_Q$ of $I\!\!F_{\tilde{q}}$ such that for all finite extensions $I\!\!F_q$ of $I\!\!F_Q$:*

$$(1) \qquad |W_3^l(I\!\!F_q)| = \frac{(1 - q^{d-1})(1 - q^d)}{(1 - q)(1 - q^2)}(1 + q)^l + q^{d-1}\frac{1 - q^d}{1 - q}.$$

In particular

$$(2) \qquad |W_3^0(I\!\!F_q)| = \frac{(1 - q^d)(1 - q^{d+1})}{(1 - q)(1 - q^2)},$$

$$|W_3^1(I\!\!F_q)| = \frac{(1 - q^d)^2}{(1 - q)^2},$$

$$|W_3^3(I\!\!F_q)| = \frac{(1 - q^d)(1 + 2q + q^2 - 3q^d - q^{d+1})}{(1 - q)^2}.$$

Proof: We have the stratification

$$\mathrm{Hilb}^3(R) = Z_{(1,1,1)} \cup Z_{(1,2)}.$$

Over the algebraic closure $\overline{I\!\!F}_q$, the stratum $Z_{(1,1,1)}$ is a fibre bundle over $\mathbf{P}_{d-1}$ with fibre $\mathbf{A}^{d-1}$. We choose the extension $I\!\!F_Q$ in such a way that the fibre bundle structure and a trivializing open cover are already defined over $I\!\!F_Q$. Now let $I\!\!F_q$ be a finite extension of $I\!\!F_Q$. Let $\mathbf{m} = (x_1, \ldots, x_d)$ be the maximal ideal in R. An ideal $I \in Z_{(1,2)}(I\!\!F_q)$ corresponds to a 2-codimensional linear subspace of $(\mathbf{m}/\mathbf{m}^2) = I\!\!F_q^d$. So we have

$$Z_{(1,2)}(I\!\!F_q) \cong Grass(2, I\!\!F_q^d).$$

An ideal $I \in Z_{(1,1,1)}(I\!\!F_q)$ is contained in a unique ideal $I' = I + \mathbf{m}^2$ of colength 2 in R. Let $I \in Z_{(1,2)}(I\!\!F_q)$. Let $f : \mathbf{m} \longrightarrow \mathbf{m}/I$ be the canonical projection. Then the ideals of colength 2 in R containing I are those of the form $f^{-1}(V)$ for $V \in \mathbf{P}(\mathfrak{m}/I)$. So we get

$$|W_3^!(\mathbb{F}_q)| = |Z_{(1,2)}(X)(\mathbb{F}_q)|(1+q)^l + |Z_{(1,1,1)}(X)(\mathbb{F}_q)|.$$

(1) follows. (2) follows from (1) by an easy computation. $\square$

From now on let $\mathbb{F}_Q$ be as in 2.5.15 and let $\mathbb{F}_q$ be a finite extension of $\mathbb{F}_Q$. Let X be a smooth projective variety over $\mathbb{F}_q$.

Definition 2.5.16. We write V_n instead of $\mathrm{Hilb}^n(R)(\mathbb{F}_q)$ and put

$$T_1 := X(\mathbb{F}_q),$$
$$T_2 := \left\{ M \subset X(\mathbb{F}_q) \,\middle|\, |M| = 2 \right\} \cup P_2(X, \mathbb{F}_q) \cup (X(\mathbb{F}_q) \times V_2),$$
$$T_3 := \left\{ M \subset X(\mathbb{F}_q) \,\middle|\, |M| = 3 \right\} \cup (X(\mathbb{F}_q) \times P_2(X, \mathbb{F}_q)) \cup P_3(X, \mathbb{F}_q)$$
$$\cup \left\{ \{x_1, (x_2, b)\} \,\middle|\, x_1 \neq x_2 \in X(\mathbb{F}_q), b \in V_2 \right\} \cup (X(\mathbb{F}_q) \times V_3).$$

Recall the notations from 2.3.6. We identify a map $f : P(X, \mathbb{F}_q) \longrightarrow V(\overline{\mathbb{F}}_q)$ with the set

$$\left\{ (\xi, I) \in P(X, \mathbb{F}_q) \times (V(\overline{\mathbb{F}}_q) \setminus V_0(\overline{\mathbb{F}}_q)) \,\middle|\, f(\xi) = I \right\}$$

and the set $M \times V_1(\overline{\mathbb{F}}_q)$ with M. In this way T_2 is identified with $T_2(X, \mathbb{F}_q)$ and T_3 with $T_3(X, \mathbb{F}_q)$ (see definition 2.3.6). Via these identifications the relation $\subset$ carries over to T_1, T_2, T_3. So by 2.3.7 there are bijections

$$\phi_1 = 1_{X(\mathbb{F}_q)} : X(\mathbb{F}_q) \longrightarrow T_1,$$
$$\phi_2 : X^{[2]}(\mathbb{F}_q) \longrightarrow T_2,$$
$$\phi_3 : X^{[3]}(\mathbb{F}_q) \longrightarrow T_3,$$

respecting $\subset$.

Lemma 2.5.17.

(1)
$$|X^{[3]}(\mathbb{F}_q)| = |X^{(3)}(\mathbb{F}_q)| + q\frac{1 - q^{d-1}}{1-q}|X(\mathbb{F}_q)|^2$$
$$+ q^2 \frac{(1 - q^{d-1})(1 - q^d)}{(1-q)(1-q^2)}|X(\mathbb{F}_q)|,$$

(2)
$$|\widetilde{\mathrm{Hilb}}^3(X)(\mathbb{F}_q)| = |(X \times X^{(2)})(\mathbb{F}_q)| + 2q\frac{1 - q^{d-1}}{1-q}|X(\mathbb{F}_q)|^2$$

$$+ q^2 \frac{(1-q^{d-1})^2}{(1-q)^2} |X(\mathbb{F}_q)|,$$

$$(3) \qquad |\widehat{H}^3(X)(\mathbb{F}_q)| = |X^3(\mathbb{F}_q)| + 3q\frac{1-q^{d-1}}{1-q}|X(\mathbb{F}_q)|^2$$

$$+ q\frac{(1-q^{d-1})(1+3q-3q^d-q^{d+1})}{(1-q)^2}|X(\mathbb{F}_q)|.$$

Proof: Immediately from the definitions we get

$$(\phi_2 \times \phi_3)(\widetilde{\mathrm{Hilb}}^3(X)(\mathbb{F}_q)) =$$

$$\left\{ (\{x_1,x_2\},\{x_1,x_2,x_3\}) \;\middle|\; \begin{array}{c} x_1,x_2,x_3 \in X(\mathbb{F}_q) \\ \text{pairwise distinct} \end{array} \right\}$$

$$\cup \left(X(\mathbb{F}_q) \times P_2(X,\mathbb{F}_q) \right)$$

$$\cup \left\{ (\{x_1,x_2\},\{(x_1,b),x_2\}) \;\middle|\; x_1 \neq x_2 \in X(\mathbb{F}_q), b \in V_2 \right\}$$

$$\cup \left\{ ((x_1,b),\{(x_1,b),x_2\}) \;\middle|\; x_1 \neq x_2 \in X(\mathbb{F}_q), b \in V_2 \right\}$$

$$\cup \left\{ ((x_1,b),(x_1,c)) \;\middle|\; x \in X(\mathbb{F}_q), b \in V_2, c \in V_3, b \supset c \right\}$$

and

$$((1_{X(\mathbb{F}_q)})^3 \times \phi_2^3 \times \phi_3)(\widehat{H}^3(X)(\mathbb{F}_q)) =$$

$$\left\{ (x_1,x_2,x_3,\{x_1,x_2\},\{x_2,x_3\},\{x_3,x_1\},\{x_1,x_2,x_3\}) \;\middle|\; \begin{array}{c} x_1,x_2,x_3 \in X(\mathbb{F}_q) \\ \text{pairwise distinct} \end{array} \right\}$$

$$\cup \left\{ (x_1,x_1,x_2,\{x_1,x_2\},\{x_1,x_2\},(x_1,b),\{(x_1,b),x_2\}) \;\middle|\; \begin{array}{c} x_1 \neq x_2 \in X(\mathbb{F}_q), \\ b \in V_2 \end{array} \right\}$$

$$\cup \left\{ (x_1,x_2,x_1,\{x_1,x_2\},(x_1,b),\{x_1,x_2\},\{(x_1,b),x_2\}) \;\middle|\; \begin{array}{c} x_1 \neq x_2 \in X(\mathbb{F}_q), \\ b \in V_2 \end{array} \right\}$$

$$\cup \left\{ (x_2,x_1,x_1,(x_1,b),\{x_1,x_2\},\{x_1,x_2\},\{(x_1,b),x_2\}) \;\middle|\; \begin{array}{c} x_1 \neq x_2 \in X(\mathbb{F}_q), \\ b \in V_2 \end{array} \right\}$$

$$\cup \left\{ (x,x,x,(x,b_1),(x,b_2),(x,b_3),(x,c)) \;\middle|\; \begin{array}{c} b_1,b_2,b_3 \subset V_2; \; c \in V_3; \\ b_1 \supset c, \; b_2 \supset c, \; b_3 \supset c \end{array} \right\}.$$

We sum the numbers of elements of T_3, $(\phi_2 \times \phi_3)(\widetilde{\mathrm{Hilb}}^3(X)(\mathbb{F}_q))$ and $(1^3_{X(\mathbb{F}_q)} \times \phi_2^3 \times \phi_3)(\widehat{H}^3(X)(\mathbb{F}_q))$ respectively. Then we use remark 1.2.4 and lemma 2.5.15 to get

$$|X^{[3]}(\mathbb{F}_q)| = \binom{|X(\mathbb{F}_q)|}{3} + |P_2(X,\mathbb{F}_q)||X(\mathbb{F}_q)| + |P_3(X,\mathbb{F}_q)|$$

$$+ \frac{1-q^d}{1-q}|X(\mathbb{F}_q)|(|X(\mathbb{F}_q)|-1)$$

$$+ \frac{(1-q^d)(1-q^{d+1})}{(1-q)(1-q^2)}|X(\mathbb{F}_q)|,$$

$$|\widetilde{\mathrm{Hilb}}^3(X)(I\!\!F_q)| = 3\binom{|X(I\!\!F_q)|}{3} + |P_2(X, I\!\!F_q)||X(I\!\!F_q)|$$

$$+ 2\frac{1-q^d}{1-q}|X(I\!\!F_q)|(|X(I\!\!F_q)| - 1) + \frac{(1-q^d)^2}{(1-q)^2}|X(I\!\!F_q)|,$$

$$|\widehat{H}^3(X)(I\!\!F_q)| = 6\binom{|X(I\!\!F_q)|}{3} + 3\frac{1-q^d}{1-q}|X(I\!\!F_q)|(|X(I\!\!F_q)| - 1)$$

$$+ \frac{(1-q^d)(1+2q+q^2-3q^d-q^{d+1})}{(1-q)^2}|X(I\!\!F_q)|.$$

By remark 1.2.4 we have

$$|X^{(3)}(I\!\!F_q)| = \binom{|X(I\!\!F_q)| + 2}{3} + |P_2(X, I\!\!F_q)||X(I\!\!F_q)| + |P_3(X, I\!\!F_q)|,$$

$$|X^{(2)}(I\!\!F_q)| = \binom{|X(I\!\!F_q)| + 1}{2} + |P_2(X I\!\!F_q)|.$$

So we get

$$|X^{[3]}(I\!\!F_q)| = |X^{(3)}(I\!\!F_q)| + \left(\frac{1-q^d}{1-q} - 1\right)|X(I\!\!F_q)|^2$$

$$+ \left(\frac{(1-q^d)(1-q^{d+1})}{(1-q)(1-q^2)} - \frac{1-q^d}{1-q}\right)|X(I\!\!F_q)|,$$

$$|\widetilde{\mathrm{Hilb}}^3(X)(I\!\!F_q)| = |X^{(2)}(I\!\!F_q)||X(I\!\!F_q)| + \left(2\frac{1-q^d}{1-q} - 2\right)|X(I\!\!F_q)|^2$$

$$+ \left(\frac{(1-q^d)^2}{(1-q)^2} + 1 - 2\frac{1-q^d}{1-q}\right)|X(I\!\!F_q)|,$$

$$|\widehat{H}^3(X)(I\!\!F_q)| = |X(I\!\!F_q)|^3 + \left(3\frac{1-q^d}{1-q} - 3\right)|X(I\!\!F_q)|^2$$

$$+ \left(\frac{(1-q^d)(1+2q+q^2-3q^d-q^{d+1})}{(1-q)^2} + 2 - 3\frac{1-q^d}{1-q}\right)|X(I\!\!F_q)|,$$

and the result follows by an easy calculation. □

Theorem 2.5.18. *Let X be a smooth projective variety over* **C**. *Then we have:*

(1)
$$p(X^{[3]}, z) = p(X^{(3)}, z) + z^2\frac{1 - z^{2d-2}}{1 - z^2}p(X, z)^2$$

$$+ z^4\frac{(1 - z^{2d-2})(1 - z^{2d})}{(1 - z^2)(1 - z^4)}p(X, z),$$

$$p(X^{[3]}, -z) = \frac{1}{6}p(X, -z)^3 + \frac{1}{2}p(X, -z^2)p(X, z) + \frac{1}{3}p(X, -z^3)$$

$$+ z^2 \frac{1 - z^{2d-2}}{1 - z^2} p(X, -z)^2$$

$$+ z^4 \frac{(1 - z^{2d-2})(1 - z^{2d})}{(1 - z^2)(1 - z^4)} p(X, -z),$$

$$(2) \qquad p(\widehat{X}^{[3]}, z) = p(X^{(3)}, z) + z^2 \frac{1 - z^{2d-2}}{1 - z^2} p(X, z)^2$$

$$+ z^2 \frac{(1 - z^{2d-2})(1 + z^2 - z^{2d} - z^{2d+2})}{(1 - z^2)^2} p(X, z),$$

$$p(\widehat{X}^{[3]}, -z) = \frac{1}{6} p(X, -z)^3 + \frac{1}{2} p(X, -z^2) p(X, z) + \frac{1}{3} p(X, -z^3)$$

$$+ z^2 \frac{1 - z^{2d-2}}{1 - z^2} p(X, -z)^2$$

$$+ z^2 \frac{(1 - z^{2d-2})(1 + z^2 - z^{2d} - z^{2d+2})}{(1 - z^2)^2} p(X, -z),$$

$$(3) \qquad p(\widetilde{\mathrm{Hilb}}^3(X), z) = p(X, z) \times p(X^{(2)}, z) + 2z^2 \frac{1 - z^{2d-2}}{1 - z^2} p(X, z)^2$$

$$+ z^4 \frac{(1 - z^{2d-2})^2}{(1 - z^2)^2} p(X, z),$$

$$p(\widetilde{\mathrm{Hilb}}^3(X), -z) = \frac{1}{2} \Big(p(X, -z)^3 + p(X, -z^2) p(X, -z) \Big)$$

$$+ 2z^2 \frac{1 - z^{2d-2}}{1 - z^2} p(X, z)^2 + z^4 \frac{(1 - z^{2d-2})^2}{(1 - z^2)^2} p(X, -z),$$

$$(4) \qquad p(\widetilde{\mathrm{Hilb}}^3(X), z) = p(X, z) \times p(X^{(2)}, z) + 2z^2 \frac{1 - z^{2d-2}}{1 - z^2} p(X, z)^2$$

$$+ z^2 \frac{(1 - z^{2d-2})(1 + 2z^2 - 2z^{2d} - z^{2d+2})}{(1 - z^2)^2} p(X, z),$$

$$p(\widetilde{\mathrm{Hilb}}^3(X), -z) = \frac{1}{2} \Big(p(X, -z)^3 + p(X, -z^2) p(X, -z) \Big)$$

$$+ 2z^2 \frac{1 - z^{2d-2}}{1 - z^2} p(X, -z)^2$$

$$+ z^2 \frac{(1 - z^{2d-2})(1 + 2z^2 - 2z^{2d} - z^{2d+2})}{(1 - z^2)^2} p(X, -z),$$

$$(5) \qquad p(\widehat{H}^3(X), z) = p(X, z)^3 + 3z^2 \frac{1 - z^{2d-2}}{1 - z^2} p(X, z)^2$$

$$+ z^2 \frac{(1 - z^{2d-2})(1 + 3z^2 - 3z^{2d} - z^{2d+2})}{(1 - z^2)^2} p(X, z).$$

Proof: X is defined over a finitely generated ring extension T of $\mathbb{Z}$, i.e. there is an X_T over $spec(T)$ satisfying $X_T \times_T \mathbb{C} = X$. Let $Y = X_T \times_T (T/m)$ be a good reduction of X modulo q. Then $Y^{[3]}$, $\widetilde{\mathrm{Hilb}}^3(Y)$ and $\widehat{H}^3(Y)$ $X^{[3]}$, $\widetilde{\mathrm{Hilb}}^3(X)$ and $\widehat{H}^3(X)$ are also reductions modulo q, and we can choose the maximal ideal $m \in spec(T)$ in such a way that they are all good reductions (see the remarks before theorem 1.2.1). Choose m in such a way that furthermore lemma 2.5.15 holds. Then (1), (3) and (5) follow immediately from lemma 2.5.17, remark 1.2.2 and Macdonald's formula. $Z_{(1,2)}(X)$ is a $Grass(2,d)$-bundle over X. So we have

$$p\big(Z_{(1,2)}(X)\big) = \frac{(1 - z^{2d-2})(1 - z^{2d})}{(1 - z^2)(1 - z^4)} p(X, z).$$

By proposition 2.5.11 we get

$$p(\widehat{X}^{[3]}, z) = p(X^{[3]}, z) + \frac{(1 - z^{2d-2})(1 - z^{2d})}{(1 - z^2)(1 - z^4)}(z^2 + z^4 + z^6)p(X, z).$$

So (2) follows from (1) by an easy computation. $B(X)$ is a $\mathbf{P}_1$-bundle over $Z_{(1,2)}(X)$. So we have by proposition 2.5.13

$$p(\widetilde{\mathrm{Hilb}}^3(X), z) = p(\widehat{\mathrm{Hilb}}^3(X), z) + (1 + z^2)\frac{(1 - z^{2d-2})(1 - z^{2d})}{(1 - z^2)(1 - z^4)}(z^2 + z^4)p(X, z).$$

(4) follows again by an easy computation. □

For a smooth projective surface S over $\mathbf{C}$ these formulas can be written as follows:

$$p(S^{[3]}, z) = p(S^{(3)}, z) + z^2 p(S, z)^2 + z^4 p(S, z)$$
$$p(\widehat{\mathrm{Hilb}}^3(S)) = p(S \times S^{(2)}, z) + 2z^2 p(S, z)^2 + z^4 p(S, z)$$
$$p(\widehat{S}^{[3]}, z) = p(S^{(3)}, z) + z^2 p(S, z)^2 + (z^2 + 2z^4 + z^6)p(S, z)$$
$$p(\widetilde{\mathrm{Hilb}}^3(S), z) = p(S \times S^{(2)}, z) + 2z^2 p(S, z)^2 + (z^2 + 3z^4 + z^6)p(S, z)$$
$$p(\widehat{H}^3(S), z) = p(S, z)^3 + 3z^2 p(S, z)^2 + (z^2 + 4z^4 + z^6)p(S, z)$$

Now we consider the case of projective space $\mathbf{P}_d$. The Chow groups $A_i(\mathbf{P}_d^{[3]})$ and $A_i(\widehat{H}^3(\mathbf{P}_d))$ have already been determined in [Rosselló-Xambo (2)].

Proposition 2.5.19. $\mathbf{P}_d^{[3]}$, $\widehat{\mathbf{P}}_d^{[3]}$, $\widetilde{\mathrm{Hilb}}^3(\mathbf{P}_d)$, $\widehat{\mathrm{Hilb}}^3(\mathbf{P}_d)$ and $\widehat{H}^3(\mathbf{P}_d)$ *all have a cell decomposition. In particular for* Y *one of these varieties* $H_{2i+1}(Y, \mathbb{Z}) = 0$; *the groups* $A_i(Y) = H_{2i}(Y, \mathbb{Z})$ *are free, and their ranks can be computed by theorem 2.5.18.*

Proof: Let $T_0, \dots, T_d$ be homogeneous coordinates on $\mathbf{P}_d$. For $i = 0, \dots n$ let p_i be the point for which $T_i = 1$ and $T_j = 0$ for $i \neq j$. Let $\Gamma \subset Sl(d+1, \mathbf{C})$ be the maximal torus of diagonal matrices and let $\lambda_0, \dots, \lambda_d$ be the linearly independent characters of Γ for which any $g \in \Gamma$ is of the form $g = diag(\lambda_0(g), \dots, \lambda_d(g))$. Then Γ acts on $\mathbf{P}_d$ by $g \cdot T_i := \lambda_i(g)T_i$. The fixed points are $p_0, \dots, p_d$. We have an induced action of Γ on $\mathbf{P}_d^{[n]}$ for all n, as Γ acts on the homogeneous ideals in $T_0, \ldots, T_d$. A subscheme $Z \in \mathbf{P}_d^{[n]}$ is a fixed point of this action, if and only if its ideal is generated by monomials in $T_0, \ldots, T_d$. So the action of Γ has only finitely many fixed points on $\mathbf{P}_d^{[n]}$. The same is true for a general one-parameter subgroup of Γ. We fix a one-parameter subgroup Φ of Γ which has only finitely many fixed points on $\mathbf{P}_d, \mathbf{P}_d^{[2]}$ and $\mathbf{P}_d^{[3]}$. The induced action of Φ on $\mathbf{P}_d^{(3)} \times (\mathbf{P}_d^{[2]})^{(3)} \times \mathbf{P}_d^{[3]}$ and $\mathbf{P}_d \times \mathbf{P}_d^{[2]} \times \mathbf{P}_d^{[3]}$ and on the quotients $\mathbf{P}_d \times \mathbf{P}_d^{(2)} \times \mathbf{P}_d^{[2]} \times (\mathbf{P}_d^{[2]})^{(2)} \times \mathbf{P}_d^{[3]}$ and $(\mathbf{P}_d)^3 \times (\mathbf{P}_d^{[2]})^3 \times \mathbf{P}_d^{[3]}$ restricts to an action on the subvarieties $\widehat{\mathbf{P}}_d^{[3]}$, $\widetilde{\mathrm{Hilb}}^3(\mathbf{P}_d)$, $\widehat{\mathrm{Hilb}}^3(\mathbf{P}_d)$ and $\widehat{H}^3(\mathbf{P}_d)$. As the action on $\mathbf{P}_d$, $\mathbf{P}_d^{[2]}$ and $\mathbf{P}_d^{[3]}$ has only finitely many fixed points, it has only finitely many on $\mathbf{P}_d \times \mathbf{P}_d^{[2]} \times \mathbf{P}_d^{[3]}$ and $(\mathbf{P}_d)^3 \times (\mathbf{P}_d^{[2]})^3 \times \mathbf{P}_d^{[3]}$. The fixed points on the quotients $\mathbf{P}_d \times \mathbf{P}_d^{(2)} \times \mathbf{P}_d^{[2]} \times (\mathbf{P}_d^{[2]})^{(2)} \times \mathbf{P}_d^{[3]}$ and $(\mathbf{P}_d)^3 \times (\mathbf{P}_d^{[2]})^3 \times \mathbf{P}_d^{[3]}$ are the images of the fixed points on $(\mathbf{P}_d)^3 \times (\mathbf{P}_d^{[2]})^3 \times \mathbf{P}_d^{[3]}$ under the quotient map. So there are also only finitely many. In particular the action of Φ has only finitely many fixed points on $\widehat{\mathbf{P}}_d^{[3]}$, $\widetilde{\mathrm{Hilb}}^3(\mathbf{P}_d)$, $\widehat{\mathrm{Hilb}}^3(\mathbf{P}_d)$ and $\widehat{H}^3(\mathbf{P}_d)$. As these are smooth, they have a cell decomposition.   □

3. The varieties of second and higher order data

The second part of this work (chapters 3 and 4) is devoted to the computation of the cohomology and Chow rings of Hilbert schemes. In chapter 3 we define varieties of second and higher order data on a smooth variety X and study them. In section 3.1 we consider the varieties $D_m^2(X)$ of second order data of m-dimensional subvarieties of X. We define $D_m^2(X)$ as a subvariety of a product of Hilbert schemes of zero-dimensional subschemes of X. Then we show that $D_m^2(X)$ can be described as a Grassmanian bundle over the Grassmannian bundle of m-dimensional subspaces of the cotangent bundle of X. $D_1^2(X)$ is a natural desingularisation of $X_{(3)}^{[3]}$. Using the description as a bundle of Grassmanians we compute the ring structure of the cohomology ring of $D_m^2(X)$. Then we descibe in what sense $D_m^2(X)$ parametrizes the second order data of m-dimensional subvarieties of X and the relation to second order contacts of such subvarieties.

In section 3.2 we consider the varieties of higher order data $D_m^n(X)$. Their definition is a generalisation of that of $D_m^2(X)$. We show that only the varieties of third order data of curves and hypersurfaces are well-behaved, i.e. they are locally trivial bundles over the corresponding varieties of second order data with fibre a projective space. In particular $D_1^3(X)$ is a natural desingularisation of $X_{(4)}^{[4]}$. Then we compute the Chow ring of these varieties. As an enumerative application of the results of chapter 3 we determine formulas for the numbers of second and third order contacts of a smooth projective variety $X \subset \mathbf{P}_N$ with linear subspaces of $\mathbf{P}_N$.

In section 3.3 we introduce the Semple bundle varieties $F_n(X)$, which parametrize higher order data of curves on X in a slightly different sense. We use them to show a general formula for the number of higher order contacts of a smooth projective variety $X \subset \mathbf{P}_N$ with lines in $\mathbf{P}_N$.

Arrondo, Sols and Speiser [Arrondo-Sols-Speiser (1)] have independently constructed new contact varieties for m-dimensional subvarieties of a given variety X, for which they also give a number of applications. Their approach is different from the one of sections 3.1 and 3.2 and is in fact a generalization of the Semple bundle construction.

This approach is more general then mine, as it gives varieties of arbitrary order. It has however the disadvantage of not taking the commutativity of higher order derivatives into account, and thus, except in the case $m = 1$, the actual data varieties are given as subvarieties (by requiring "symmetry") of considerably bigger varieties. The precise description of these subvarieties appears to be not a very easy task, and as far as I know has been carried out only in the case of second order data of surfaces in $\mathbf{P}_3$.

3.1. The varieties of second order data.

Let X be a smooth projective variety of dimension d over an algebraically closed field k. In this section we want to define a variety $D_m^2(X)$ of second order data of m-dimensional subvarieties of X for any non-negative integer $m \leq d$. A general point of $D_m^2(X)$ will correspond to the second order datum of the germ of a smooth m-dimensional subvariety $Y \subset X$ in a point $x \in X$, i.e. to the quotient $\mathcal{O}_{Y,x}/\mathbf{m}_{X,x}^3$ of $\mathcal{O}_{X,x}$. Assume for the moment that the ground field is $\mathbf{C}$ and $x \in Y \subset X$, X is a smooth complex d-manifold and we have local coordinates $z_1, \ldots, z_d$ at x. Then Y is given by equations

$$f_i(z_1, \ldots, z_d) = 0 \quad i = 1, \ldots, d - m.$$

Then the second order datum $\mathcal{O}_{Y,x}/\mathbf{m}_{X,x}^3$ is

$$\mathbf{C}[z_1, \ldots, z_d]/((f_1, \ldots, f_{d-m}) + \mathbf{m}^3),$$

and giving the second order datum is equivalent to giving the derivatives

$$\frac{\partial f_i}{\partial z_j}(x), \ i = 1, \ldots, d-m, \ j = 1, \ldots, d$$

$$\frac{\partial^2 f_i}{\partial z_j \partial z_l}(x), \ i = 1, \ldots, d-m, \ j, l = 1, \ldots, d$$

Notation. In chapter 3 and 4 we will often use the Grassmannian bundle associated to a vector bundle. So we fix some notations for these.

Let S be a scheme and E a vector bundle of rank r on X. For any $m \leq r$ let $Grass(m, E)$ denote the Grassmannian bundle of m-dimensional quotients of E. Let $\pi_{m,E} : Grass(m, E) \longrightarrow S$ be the projection, $Q_{m,E}$ the universal quotient bundle of $\pi_{m,E}^*(E)$ and $T_{r-m,E}$ the tautological subbundle. Then the projectivization of E is $\mathbf{P}(E) = Grass(r-1, E)$ and $\mathcal{O}_{\mathbf{P}(E)}(1) = (T_{1,E})^*$. We also put $\check{\mathbf{P}}(E) := Grass(1, E)$. We write $Grass(m, r)$ for the Grassmann variety of m-dimensional quotients of $\mathbf{C}^r$. Let $Q_{m,r}$ and $T_{r-m,r}$ be the universal quotient bundle and the tautological subbundle on $Grass(m, r)$.

Notation. For subschemes Z_1, Z_2 of a scheme S with ideal sheaves $\mathcal{I}_{Z_1}$, $\mathcal{I}_{Z_2}$ respectively in $\mathcal{O}_S$, let $Z_1 \cdot Z_2$ denote the subscheme Z of S whose ideal sheaf $\mathcal{I}_Z$ is given by $\mathcal{I}_Z := \mathcal{I}_{Z_1} \cdot \mathcal{I}_{Z_2}$.

As above we will write $Z_1 \subset Z_2$; to mean that Z_1 is a subscheme of Z_2. In this case we will write $\mathcal{I}_{Z_1/Z_2}$ for the ideal of Z_1 in Z_2.

Definition 3.1.1. Let $\mathcal{D}_m^2(X)$ be the contravariant functor from the category of noetherian k-schemes to the category of sets which for noetherian k-schemes S, T and a morphism $\phi : S \longrightarrow T$ is given by:

$$
\mathcal{D}_m^2(X)(T) = \left\{ (Z_0, Z_1, Z_2) \;\middle|\; \begin{array}{c} Z_0, Z_1, Z_2 \subset X \times T \\ \text{closed subschemes} \\ \text{flat of degrees } 1, m+1, \binom{m+2}{2} \text{ over } T \\ Z_0 \subset Z_1 \subset Z_2, \; Z_1 \subset Z_0 \cdot Z_0, \\ Z_2 \subset Z_0 \cdot Z_1 \end{array} \right\},
$$

$$
\begin{array}{lccc}
\mathcal{D}_m^2(X)(\phi): & \mathcal{D}_m^2(X)(T) & \longrightarrow & \mathcal{D}_m^2(X)(S) \\
& (Z_0, Z_1, Z_2) & \longmapsto & (Z_0 \times_T S, Z_1 \times_T S, Z_2 \times_T S).
\end{array}
$$

Lemma 3.1.2. $\mathcal{D}_m^2(X)$ *is representable by a closed subscheme* $D_m^2(X) \subset X \times X^{[m+1]} \times X^{\left[\binom{m+2}{2}\right]}$.

Proof: Let

$$
\begin{aligned}
Z_1(X) &:= \Delta \subset X \times X \\
Z_{m+1}(X) &\subset X \times X^{[m+1]} \\
Z_{\binom{m+2}{2}}(X) &\subset X \times X^{\binom{m+2}{2}}
\end{aligned}
$$

be the universal subschemes. To shorten notations we write

$$
W := X \times X^{[m+1]} \times X^{\left[\binom{m+2}{2}\right]}.
$$

For $i = 1, 2, 3$ let p_i be the projection of W to the ith factor. Let $\mathcal{I}_0, \mathcal{I}_1, \mathcal{I}_2$ be the ideals of

$$
\begin{aligned}
W_0 &:= (1_X \times p_1)^{-1}(Z_1(X)), \\
W_1 &:= (1_X \times p_2)^{-1}(Z_{m+1}(X)), \\
W_2 &:= (1_X \times p_3)^{-1}(Z_{\binom{m+2}{2}}(X))
\end{aligned}
$$

respectively in $\mathcal{O}_{X \times W}$. Let $U_0, U_1, U_2 \subset W$ be the subschemes defined by $\mathcal{I}_0 + \mathcal{I}_1$, $\mathcal{I}_1 + \mathcal{I}_2 + \mathcal{I}_0^2$ und $\mathcal{I}_2 + \mathcal{I}_0 \cdot \mathcal{I}_1$ respectively. Then we have obviously $U_i \subset W_i$ for $i = 0, 1, 2$. As X is a closed subvariety of a projective space $\mathbf{P}_N$, W_0, W_1, W_2, U_0, U_1, U_2 are in a natural way subschemes of $\mathbf{P}_N \times W$. The W_i are flat of degree $\binom{i+m}{i}$ over W for $i = 0, 1, 2$. We put

$$
\mathcal{F} := \mathcal{O}_{U_0} \oplus \mathcal{O}_{U_1} \oplus \mathcal{O}_{U_2}.
$$

For any morphism $g : T \longrightarrow W$ of a noetherian scheme to W we put

$$
\mathcal{F}_g := (1_{\mathbf{P}_N} \times g)^*(\mathcal{F})
$$

on $\mathbf{P}_N \times T$. Let $\pi_T : \mathbf{P}_N \times T \longrightarrow T$ be the projection. By [Mumford (1) Lecture 8] there is a closed subscheme $D_m^2(X) \subset W$ such that the following holds: $(\pi_T)_* \mathcal{F}_g$ is locally free of rank

$$r_1 := 1 + m + 1 + \binom{m+2}{2}$$

over T if and only if g factors through $D_m^2(X)$. $(D_m^2(X)$ is closed and not only locally closed as each U_i is a subscheme of the corresponding W_i, and so $(\pi_T)_*(\mathcal{F}_g)$ can at most have rank r_1 in points of T.) By the relations $U_0 \subset W_0$, $U_1 \subset W_1$, $U_2 \subset W_2$, $(\pi_T)_*(\mathcal{F}_g)$ is locally free of rank r_1 if and only if

$$(1_{\mathbf{P}_N} \times g)^{-1}(U_i) = (1_{\mathbf{P}_N} \times g)^{-1}(W_i) \qquad i = 0, 1, 2.$$

Now we can easily see from the definitions that $D_m^2(X)$ represents the functor $\mathcal{D}_m^2(X)$.   □

Definition 3.1.3. Let

$$D_m^2(X) \subset X \times X^{[m+1]} \times X^{\left[\binom{m+2}{2}\right]}$$

be the subscheme representing the functor $\mathcal{D}_m^2(X)$ by lemma 3.1.2. As a set it is given by

$$\left\{ \begin{array}{c} (x, Z_1, Z_2) \\ \in X \times X^{[m+1]} \times X^{\left[\binom{m+2}{2}\right]} \end{array} \middle| \begin{array}{c} x \subset Z_1 \subset Z_2, \\ Z_1 \subset x \cdot x, \\ Z_2 \subset x \cdot Z_1 \end{array} \right\}.$$

Later we will see that $D_m^2(X)$ is reduced and even smooth. $D_m^2(X)$ is called the variety of second order data of m-dimensional subvarieties of X. Analogously we define $D_m^1(X)$ as the closed subscheme of $X \times X^{[m+1]}$ that represents the functor given by

$$\mathcal{D}_m^1(X)(T) := \left\{ (Z_0, Z_1) \middle| \begin{array}{c} Z_0, Z_1 \subset X \times T \text{ closed subschemes} \\ \text{flat of degree } 1, m+1 \text{ respectively over } T \\ Z_0 \subset Z_1 \subset Z_0 \cdot Z_0 \end{array} \right\}.$$

$D_m^1(X)$ is the variety of first order data of m-dimensional subschemes of X. As a set it is obviously given by

$$\left\{ (x, Z) \in X \times X^{[m+1]} \middle| x \subset Z \subset x \cdot x \right\}.$$

We will also see that $D_m^1(X)$ is smooth.

For a surface S the variety $D_1^2(S)$ is considered in the literature (using a slightly different definition). It is called the variety of second order data on S and denoted by $D(S)$. $D_1^2(\mathbf{P}_2)$ was studied extensively in [Roberts-Speiser (1),(2),(3),(4)] and [Roberts (1)] to find enumerative formulas for second order contacts of families of curves in $\mathbf{P}_2$. For a surface S the variety $D_1^2(S)$ has been studied in [Collino (1)], and there its cohomology ring was determined. In [Le Barz (10)] $D_1^2(X)$ has been defined for a general smooth projective variety X over $\mathbf{C}$ as a subvariety of $\widehat{H}^3(X)$ (see section 2.5).

We now give another definition $\widetilde{D}_m^2(X)$ of $D_m^2(X)$, which will enable us to compute the Chow ring of this variety. We then have to show that $D_m^2(X)$ and $\widetilde{D}_m^2(X)$ are isomorphic.

Definition 3.1.4. Let again $J_n(X)$ be the nth jet-bundle of X. Let $\tilde{\pi}_1 : Grass(m, T_X^*) \longrightarrow X$ be the projection and $T_1 := T_{d-m,T_X^*}$, $Q_1 := Q_{m,T_X^*}$. We also write $\widetilde{D}_m^1(X)$ for $Grass(m, T_X^*)$. Let $j_1 : \tilde{\pi}_1^*(T_X^*) \longrightarrow \tilde{\pi}_1^*(J_1(X))$ be the canonical inclusion and

$$0 \longrightarrow T_1 \xrightarrow{i_1} \tilde{\pi}_1^*(T_X^*) \xrightarrow{q_1} Q_1 \longrightarrow 0$$

the canonical exact sequence. We define the vector bundle $\widetilde{Q}_1$ on $\widetilde{D}_m^1(X)$ by the following commutative diagram with exact rows and columns.

$$
\begin{array}{ccccccccc}
 & & 0 & & 0 & & 0 & & \\
 & & \downarrow & & \downarrow & & \downarrow & & \\
0 & \longrightarrow & T_1 & \xrightarrow{i_1} & \tilde{\pi}_1^*(T_X^*) & \xrightarrow{q_1} & Q_1 & \longrightarrow & 0 \\
 & & \| & & \downarrow{\scriptstyle j_1} & & \downarrow{\scriptstyle \bar{j}_1} & & \\
0 & \longrightarrow & T_1 & \xrightarrow{i} & \tilde{\pi}_1^*(J_1(X)) & \xrightarrow{\tilde{q}_1} & \widetilde{Q}_1 & \longrightarrow & 0 \\
 & & \downarrow & & \downarrow & & \downarrow{\scriptstyle \bar{q}_1} & & \\
0 & \longrightarrow & \mathcal{O}_{\widetilde{D}_m^1(X)} & = & \mathcal{O}_{\widetilde{D}_m^1(X)} & \longrightarrow & 0 & & \\
 & & \downarrow & & \downarrow & & & & \\
 & & 0 & & 0 & & & &
\end{array}
$$

Let $s^2(q_1) : \mathrm{Sym}^2(\tilde{\pi}_1^*(T_X^*)) \longrightarrow \mathrm{Sym}^2(Q_1)$ be the morphism induced by the quotient morphism $q_1 : \tilde{\pi}_1^*(T_X^*) \longrightarrow Q_1$, and let $j_2 : \tilde{\pi}_1^*(\mathrm{Sym}^2(T_X^*)) \longrightarrow \tilde{\pi}_1^*(J_2(X))$ be the canonical map.

We define the vector bundle $\widetilde{T}_1$ on $Grass(m, T_X^*)$ by the following diagram with exact rows and columns in which the right lower square is cartesian

$$
\begin{array}{ccccccc}
 & & 0 & & 0 & & \\
 & & \uparrow & & \uparrow & & \\
0 & \longrightarrow & \tilde{Q}_1 & = & \tilde{Q}_1 & \longrightarrow & 0 \\
 & & \uparrow & & \uparrow \scriptstyle{\tilde{q}_1} & & \\
0 \longrightarrow \tilde{\pi}_1^*(\mathrm{Sym}^2(T_X^*)) & \xrightarrow{\tilde{j}_2} & \tilde{\pi}_1^*(J_2(X)) & \longrightarrow & \tilde{\pi}_1^*(J_1(X)) & \longrightarrow & 0 \\
\| & & \uparrow \scriptstyle{\tilde{i}_1} \quad \square & & \uparrow \scriptstyle{i} & & \\
0 \longrightarrow \mathrm{Sym}^2(\tilde{\pi}_1^*(T_X^*)) & \xrightarrow{\tilde{j}_2} & \tilde{T}_1 & \longrightarrow & T_1 & \longrightarrow & 0 \\
 & & \uparrow & & \uparrow & & \uparrow \\
 & & 0 & & 0 & & 0
\end{array}
$$

and $W_m^2(X)$ and $(T_1 \cdot T_X^*)$ by the following diagram with exact rows and columns in which the left lower square is cartesian

$$
\begin{array}{ccccccc}
 & & 0 & & 0 & & \\
 & & \downarrow & & \downarrow & & \\
0 & \longrightarrow & (T_1 \cdot T_X^*) & = & (T_1 \cdot T_X^*) & \longrightarrow & 0 \\
 & & \downarrow & & \downarrow & & \downarrow \\
0 & \longrightarrow & \tilde{\pi}_1^*(\mathrm{Sym}^2(T_X^*)) & \xrightarrow{\tilde{j}_2} & \tilde{T}_1 & \longrightarrow & T_1 \longrightarrow 0 \\
 & & \downarrow \scriptstyle{s^2(q_1)} \quad \square & & \downarrow \scriptstyle{p} & & \| \\
0 & \longrightarrow & \mathrm{Sym}^2(Q_1) & \xrightarrow{\hat{j}_2} & W_m^2(X) & \xrightarrow{\hat{q}_2} & T_1 \longrightarrow 0 \\
 & & \downarrow & & \downarrow & & \downarrow \\
 & & 0 & & 0 & & 0
\end{array}
$$

Obviously $W_m^2(X)$ is a vector bundle of rank

$$
r = \binom{m+1}{2} + d - m
$$

over $Grass(m, T_X^*)$. $(T_1 \cdot T_X^*)$ is also the image of the subbundle $T_1 \otimes \tilde{\pi}_1^*(T_X^*)$ under the natural vector bundle morphism $s_2 : \tilde{\pi}_1^*(T_X^* \otimes T_X^*) \longrightarrow \tilde{\pi}_1^*(\mathrm{Sym}^2(T_X^*))$. We can see easily that $(T_1 \cdot T_X^*)$ is a vector bundle, and from the diagram we get

$$
\mathrm{Sym}^2(Q_1) = \mathrm{Sym}^2(T_X^*)/(T_1 \cdot T_X^*).
$$

$s^2(q_1) : \operatorname{Sym}^2(T_X^*) \longrightarrow \operatorname{Sym}^2(Q_1)$ is the quotient map.

Definition 3.1.5. We put

$$\tilde{D}_m^2(X) := Grass\big((\tbinom{m+1}{2}), W_m^2(X)\big).$$

Let $\tilde{\pi}_2 := \tilde{D}_m^2(X) \longrightarrow Grass(m, T_X^*)$ be the projection. Let $T_2 := T_{d-m, W_m^2(X)}$. Let

$$0 \longrightarrow T_2 \xrightarrow{\ i_2\ } \tilde{\pi}_2^*(W_m^2(X)) \xrightarrow{\ q_2\ } Q_2 \longrightarrow 0$$

be the natural exact sequence. We define the vector bundle $\tilde{T}_2$ on $\tilde{D}_m^2(X)$ by the following diagram with exact rows and columns in which the upper right square is cartesian.

$$
\begin{array}{ccccccccc}
 & & 0 & & 0 & & 0 & & \\
 & & \uparrow & & \uparrow & & \uparrow & & \\
0 & \longrightarrow & T_2 & \xrightarrow{i_2} & \tilde{\pi}_2^*(W_m^2(X)) & \xrightarrow{q_2} & Q_2 & \longrightarrow & 0 \\
 & & \uparrow & \square & \uparrow {\scriptstyle \tilde{\pi}_2^*(p)} & & \| & & \\
0 & \longrightarrow & \tilde{T}_2 & \xrightarrow{\tilde{i}_2} & \tilde{\pi}_2^*(\tilde{T}_1) & \xrightarrow{\tilde{q}_2} & Q_2 & \longrightarrow & 0 \\
 & & \uparrow & & \uparrow & & \uparrow & & \\
0 & \longrightarrow & \tilde{\pi}_2^*(T_1 \cdot T_X^*) & = & \tilde{\pi}_2^*(T_1 \cdot T_X^*) & \longrightarrow & 0 & & \\
 & & \uparrow & & \uparrow & & & & \\
 & & 0 & & 0 & & & &
\end{array}
$$

The vector bundle $\tilde{Q}_2$ is now defined by the following diagram :

$$
\begin{array}{ccccccccc}
 & & 0 & & 0 & & 0 & & \\
 & & \downarrow & & \downarrow & & \downarrow & & \\
0 & \longrightarrow & \tilde{T}_2 & \xrightarrow{\tilde{i}_2} & \tilde{\pi}_2^*(\tilde{T}_1) & \xrightarrow{\tilde{q}_2} & Q_2 & \longrightarrow & 0 \\
 & & \| & & \downarrow {\scriptstyle \tilde{\pi}_2^*(\tilde{i}_1)} & & \downarrow {\scriptstyle \tilde{j}_2} & & \\
0 & \longrightarrow & \tilde{T}_2 & \longrightarrow & \tilde{\pi}_2^*(\tilde{\pi}_1^*(J_2(X))) & \longrightarrow & \tilde{Q}_2 & \longrightarrow & 0 \\
 & & \downarrow & & \downarrow & & \downarrow {\scriptstyle \bar{q}_2} & & \\
0 & \longrightarrow & & & \tilde{\pi}_2^*(\tilde{Q}_1) & = & \tilde{\pi}_2^*(\tilde{Q}_1) & \longrightarrow & 0 \\
 & & & & \downarrow & & \downarrow & & \\
 & & & & 0 & & 0 & &
\end{array}
$$

From these diagrams we can read off the exact sequences

$$0 \longrightarrow \operatorname{Sym}^2(Q_1) \xrightarrow{\hat{j}_2} W_m^2(X) \xrightarrow{\hat{q}_2} T_1 \longrightarrow 0$$

and

$$0 \longrightarrow Q_1 \xrightarrow{\tilde{j}_1} \tilde{Q}_1 \xrightarrow{\tilde{q}_1} \mathcal{O}_{\tilde{D}_m^1(X)} \longrightarrow 0$$

on $D_m^1(X)$ and

$$0 \longrightarrow Q_2 \xrightarrow{\tilde{j}_2} \tilde{Q}_2 \xrightarrow{\tilde{q}_2} \tilde{\pi}_2^*(\tilde{Q}_1) \longrightarrow 0$$

on $D_m^2(X)$.

Definition 3.1.6. For any $n \in \mathbb{N}$ let as above $Z_n(X) \subset X \times X^{[n]}$ be the universal subscheme with the projections:

$$
\begin{array}{ccc}
 & Z_n(X) & \\
\swarrow{\scriptstyle p_n} & & \searrow{\scriptstyle q_n} \\
X & & X^{[n]}.
\end{array}
$$

Let

$$r_1 : D_m^1(X) \longrightarrow X^{[m+1]},$$
$$r_2 : D_m^2(X) \longrightarrow X^{\left[\binom{m+2}{2}\right]}$$

be the projections. We put

$$(\mathcal{O}_X)_m^1 := r_1^*(q_{m+1})_*(\mathcal{O}_{Z_{m+1}(X)}) = r_1^*(q_{m+1})_* p_{m+1}^*(\mathcal{O}_X),$$
$$(\mathcal{O}_X)_m^2 := r_2^*(q_{\binom{m+2}{2}})_*(\mathcal{O}_{Z_{\binom{m+2}{2}}(X)}) = r_2^*(q_{\binom{m+2}{2}})_* p_{\binom{m+2}{2}}^*(\mathcal{O}_X).$$

$(\mathcal{O}_X)_m^1$ is a vector bundle of rank $m+1$ on $D_m^1(X)$ and $(\mathcal{O}_X)_m^2$ a vector bundle of rank $\binom{m+2}{2}$ on $D_m^2(X)$. Let $\Delta \subset X \times X$ be the diagonal and $\mathcal{I}_\Delta \subset \mathcal{O}_{X \times X}$ its ideal sheaf. For all $n \in \mathbb{N}$ let Δ^n be the subscheme of $X \times X$ defined by $(\mathcal{I}_\Delta)^n$ (which has support $\Delta \cong X$). Let $s_1, s_2 : X \times X \longrightarrow X$ be the projections. Then we have

$$J_n(X) = (s_2)_*(\mathcal{O}_{X \times X}/(\mathcal{I}_\Delta)^{n+1}) = (s_2)_*(\mathcal{O}_{\Delta^{n+1}}).$$

Let $\phi : T \longrightarrow X$ be a morphism from a noetherian scheme. We define $\Delta_\phi \subset X \times T$ by

$$
\begin{array}{ccc}
\Delta & \longrightarrow & X \times X \\
\uparrow & \square & \uparrow{\scriptstyle 1_X \times \phi} \\
\Delta_\phi & \longrightarrow & X \times T.
\end{array}
$$

Then the projection $p_T : \Delta_\phi \longrightarrow T$ is an isomorphism. Analogously we define for all $n \in I\!N$ the subscheme $\Delta_\phi^n \subset X \times T$ by

$$
\begin{array}{ccc}
\Delta^n & \longrightarrow & X \times X \\
\uparrow & \square & \uparrow{\scriptstyle 1_X \times \phi} \\
\Delta_\phi^n & \longrightarrow & X \times T
\end{array}
$$

Theorem 3.1.7. *There exist isomorphisms*

$$
\phi_1 : D_m^1(X) \longrightarrow Grass(m, T_X^*),
$$
$$
\phi_2 : D_m^2(X) \longrightarrow \widetilde{D}_m^2(X),
$$

for which the diagram

$$
\begin{array}{ccc}
D_m^2(X) & \xrightarrow{\ \phi_2\ } & \widetilde{D}_m^2(X) \\
\downarrow{\scriptstyle \pi_2} & & \downarrow{\scriptstyle \tilde{\pi}_2} \\
D_m^1(X) & \xrightarrow{\ \phi_1\ } & Grass(m, T_X^*) \\
\ \ \searrow{\scriptstyle \pi_1} & & \downarrow{\scriptstyle \tilde{\pi}_1} \\
& X &
\end{array}
$$

commutes such that

$$
\phi_1^*(\widetilde{Q}_1) = (\mathcal{O}_X)_m^1,
$$
$$
\phi_2^*(\widetilde{Q}_2) = (\mathcal{O}_X)_m^2.
$$

Proof: With the notations of definition 3.1.6 we can rewrite the functors $\mathcal{D}_m^1(X)$, $\mathcal{D}_m^2(X)$ as:

$$
\mathcal{D}_m^1(X)(T) := \left\{ (\phi, Z_1) \ \middle| \
\begin{array}{c}
\phi : T \longrightarrow X \\
Z_1 \subset X \times T \text{ closed subscheme} \\
\text{flat of degree } m+1 \text{ over } T \text{ with} \\
\Delta_\phi \subset Z_1 \subset \Delta_\phi^2
\end{array}
\right\},
$$

$$
\mathcal{D}_m^2(X)(T) := \left\{ (\phi, Z_1, Z_2) \ \middle| \
\begin{array}{c}
\phi : T \longrightarrow X \\
Z_1, Z_2 \subset X \times T \text{ closed subschemes} \\
\text{flat over } T \text{ of degrees } m+1 \\
\text{and } \binom{m+2}{2} \text{ respectively with} \\
\Delta_\phi \subset Z_1 \subset \Delta_\phi^2; \\
Z_1 \subset Z_2 \subset \Delta_\phi \cdot Z_1
\end{array}
\right\}.
$$

Let $Z_1 \subset X \times D_m^1(X)$ be the universal family of subschemes flat of degree $m+1$ over $D_m^1(X)$. Then we have

$$\Delta_{\pi_1} \subset Z_1 \subset \Delta_{\pi_1}^2.$$

Let $q_1 : \Delta_{\pi_1} \longrightarrow D_m^1(X)$ be the projection. $(q_1)_*(\mathcal{I}_{\Delta_{\pi_1}/Z_1})$ is a locally free quotient of $(q_1)_*(\mathcal{I}_{\Delta_{\pi_1}/\Delta_{\pi_1}^2}) = \pi_1^*(T_X^*)$ of rank m. This defines a morphism

$$\phi_1 : D_m^1(X) \longrightarrow Grass(m, T_X^*)$$

over X.

We get the inverse as follows: for the variety $\Delta_{\tilde{\pi}_1} \subset X \times Grass(m, T_X^*)$ the projection p_1 to $Grass(m, T_X^*)$ is an isomorphism, and we have

$$\tilde{\pi}_1^*(J_n(X)) = (p_1)_*(\mathcal{O}_{\Delta_{\tilde{\pi}_1}^{n+1}}).$$

The quotient $\tilde{Q}_1$ of $\tilde{\pi}_1^*(J_1(X)) = (p_1)_*(\mathcal{O}_{\Delta_{\tilde{\pi}_1}^2})$ defines a subscheme $Z_1 \subset \Delta_{\tilde{\pi}_1}^2$ satisfying $\Delta_{\tilde{\pi}_1} \subset Z_1$. The pair $(\tilde{\pi}_1, Z_1)$ defines the required morphism

$$\psi_1 : Grass(m, T_X^*) \longrightarrow D_m^1(X)$$

over X. We see that ψ_1 is the inverse of ϕ_1.

To construct ϕ_2, ϕ_2^{-1} we proceed in a similar way.

Let $Z_1, Z_2 \subset X \times D_m^2(X)$ be the universal subschemes of degrees $m+1$, $\binom{m+2}{2}$ over $D_m^2(X)$. Via ϕ_1 we identify $Grass(m, T_X^*)$ with $D_m^1(X)$. By definition we have

$$\Delta_{\pi_1 \circ \pi_2} \subset Z_1 \subset \Delta_{\pi_1 \circ \pi_2}^2,$$
$$Z_1 \subset Z_2 \subset \Delta_{\pi_1 \circ \pi_2} \cdot Z_1 \subset \Delta_{\pi_1 \circ \pi_2}^3.$$

Let q_1 be the projection of $\Delta_{\pi_1 \circ \pi_2}$ to $D_m^2(X)$. $(q_1)_*(\mathcal{I}_{Z_1/Z_2})$ is a locally free quotient of

$$(q_1)_*(\mathcal{I}_{Z_1/\Delta_{\pi_1 \circ \pi_2} \cdot Z_1}) = \pi_2^*(W_m^2(X))$$

of rank $\binom{m+1}{2}$. This defines a morphism $\phi_2 : D_m^2(X) \longrightarrow \tilde{D}_m^2(X)$ over $D_m^1(X)$.

Let $Z_1 := \tilde{\pi}_2^{-1}(W_1)$ where W_1 is the universal subscheme over $D_m^2(X)$ of degree $m+1$. $\Delta_{\tilde{\pi}_1 \circ \tilde{\pi}_2} \subset X \times \tilde{D}_m^2(X)$ is via the projection to the second factor isomorphic to $\tilde{D}_m^2(X)$. We have

$$\tilde{\pi}_2^*(\tilde{\pi}_1^*(J_n(X))) = (p_2)_*(\mathcal{O}_{\Delta_{\tilde{\pi}_1 \circ \tilde{\pi}_2}^{n+1}}).$$

$\tilde{T}_2$ is a subbundle of $\tilde{\pi}_2^*(\tilde{T}_1)$ and $\tilde{\pi}_2^*(\tilde{T}_1)$ is a subbundle of $(p_2)_*(\mathcal{O}_{\Delta_{\tilde{\pi}_1 \circ \tilde{\pi}_2}^3})$. By the definitions $\tilde{\pi}_2^*(T_1 \cdot T_X^*)$ is a subbundle of $\tilde{T}_2$. Let $I_2 \subset \mathcal{O}_{\Delta_{\tilde{\pi}_1 \circ \tilde{\pi}_2}^3}$ be the $\mathcal{O}_{\Delta_{\tilde{\pi}_1 \circ \tilde{\pi}_2}}$-submodule with $(p_2)_*(I_2) = \tilde{T}_2$. As $\tilde{T}_2$ is a subbundle of $\tilde{\pi}_2^*(\tilde{T}_1)$, we have $I_2 \subset \mathcal{I}_{Z_1/\Delta_{\tilde{\pi}_1 \circ \tilde{\pi}_2}^3}$. As $\tilde{\pi}_2^*(T_1 \cdot T_X^*)$ is a subbundle of $\tilde{T}_2$, we have

$$\mathcal{I}_{Z_1/\Delta_{\tilde{\pi}_1 \circ \tilde{\pi}_2}^3} \cdot \mathcal{I}_{\Delta_{\tilde{\pi}_1 \circ \tilde{\pi}_2}/\Delta_{\tilde{\pi}_1 \circ \tilde{\pi}_2}^3} \subset I_2.$$

So we have in particular

$$\mathcal{O}_{\Delta^3_{\tilde{\pi}_1 \circ \tilde{\pi}_2}} \cdot I_2 \subset I_2.$$

So I_2 is an ideal in $\mathcal{O}_{\Delta^3_{\tilde{\pi}_1 \circ \tilde{\pi}_2}}$, and thus defines a subscheme $Z_2 \subset \Delta^3_{\tilde{\pi}_1 \circ \tilde{\pi}_2}$. By $I_2 \subset \mathcal{I}_{Z_1/\Delta^3_{\tilde{\pi}_1 \circ \tilde{\pi}_2}}$ we have $Z_1 \subset Z_2$ and by

$$\mathcal{I}_{Z_1/\Delta^3_{\tilde{\pi}_1 \circ \tilde{\pi}_2}} \cdot \mathcal{I}_{\Delta_{\tilde{\pi}_1 \circ \tilde{\pi}_2}/\Delta^3_{\tilde{\pi}_1 \circ \tilde{\pi}_2}} \subset I_2$$

we get $Z_2 \subset \Delta_{\tilde{\pi}_1 \circ \tilde{\pi}_2} \cdot Z_1$. The triple $(\tilde{\pi}_1 \circ \tilde{\pi}_2, Z_1, Z_2)$ defines the morphism $\psi_2 : \tilde{D}^2_m(X) \longrightarrow D^2_m(X)$ over ϕ_1 satisfying

$$\psi_2^*((\mathcal{O}_X)^2_m) = (\tilde{\pi}_2^* \tilde{\pi}_1^*(J_2(X)))/\tilde{T}_2 = \tilde{Q}_2.$$

Obviously we have $\psi_2 = \phi_2^{-1}$. $\quad\square$

In future we want to identify $\tilde{D}^1_m(X)$ with $D^1_m(X)$ and $\tilde{D}^2_m(X)$ with $D^2_m(X)$ via ϕ_1 and ϕ_2.

Remark 3.1.8.

(1) The closure of $Z_{\left(1,m,\binom{m+1}{2}\right)}(X)$ in $X^{\left[\binom{m+2}{2}\right]}$ is

$$\overline{Z}_{\left(1,m,\binom{m+1}{2}\right)}(X) = \left\{ Z \in X^{\left[\binom{m+2}{2}\right]} \;\middle|\; \begin{array}{c} \text{there are } x \in X, \, Z_1 \in Z_{(1,m)}(X) \\ \text{with } x \subset Z_1 \subset Z \subset x \cdot Z_1 \end{array} \right\}.$$

(2) The projection $r_2 : D^2_m(X) \longrightarrow \overline{Z}_{\left(1,m,\binom{m+1}{2}\right)}(X)$ is a natural resolution of $\overline{Z}_{\left(1,m,\binom{m+1}{2}\right)}(X)$.

(3) $r_2 : D^2_1(X) \longrightarrow X^{[3]}_{(3)}$ is a natural resolution of $X^{[3]}_{(3)}$. It is the blow up along $Z_{(1,2)}(X)$.

Proof: $D^2_m(X)$ is closed and irreducible. By the definitions $\overline{Z}_{\left(1,m,\binom{m+1}{2}\right)}(X)$ is the image of the projection $r_2 : D^2_m(X) \longrightarrow X^{\left[\binom{m+2}{2}\right]}$. As $D^2_m(X)$ and $Z_{\left(1,m,\binom{m+1}{2}\right)}(X)$ are smooth, we can easily see that r_2 is an isomorphism over the open subset $Z_{\left(1,m,\binom{m+1}{2}\right)}(X)$ of $\overline{Z}_{\left(1,m,\binom{m+1}{2}\right)}(X)$. $r_2^{-1}(Z_{\left(1,m,\binom{m+1}{2}\right)})(X)$ is dense in $D^2_m(X)$, as $D^2_m(X)$ is irreducible. So $\overline{Z}_{\left(1,m,\binom{m+1}{2}\right)}(X)$ is the closure of $Z_{\left(1,m,\binom{m+1}{2}\right)}(X)$ in $X^{\left[\binom{m+2}{2}\right]}$. As $D^2_m(X)$ is smooth, it is a resolution of $\overline{Z}_{\left(1,m,\binom{m+1}{2}\right)}(X)$. It is easy to see that $X^{[3]}_{(3)}$ is the closure of $Z_{(1,1,1)}(X)$. $r_2 : D^2_1(X) \longrightarrow X^{[3]}_{(3)}$ is an isomorphism over $Z_{(1,1,1)}(X)$. $Z_{(1,2)}(X)$ has codimension 2 in $X^{[3]}_{(3)}$, as

$$X^{[3]}_{(3)} = Z_{(1,1,1)}(X) \cup Z_{(1,2)}(X),$$

and $Z_{(1,1,1)}(X)$ is an $\mathbf{A}^{d-1}$-bundle over $\check{\mathbf{P}}(T_X^*)$ and $Z_{(1,2)}(X) = Grass(2, T_X^*)$. We have the exact sequence

$$0 \longrightarrow Q_1^{\otimes 2} \xrightarrow{\check{j}_2} W_1^2(X) \xrightarrow{\check{q}_2} T_1 \longrightarrow 0.$$

Let

$$D_1^2(X)_\infty := \check{\mathbf{P}}(T_1) \subset \check{\mathbf{P}}(W_1^2(X)) = D_1^2(X).$$

We see that $D_2^1(X)_\infty$ is an irreducible divisor in $D_1^2(X)$ and

$$(D_2^1(X)_\infty) = r_2^{-1}(Z_{(1,2)}(X)).$$

(3) follows with the same argument as in the end of the proof of proposition 2.5.8.
□

For the rest of section 3.1 let d_i, e_i, f_i, g_i be variables of weight i. Each class $b \in A^i(X)$ will also be given weigth i. Let E be a vector bundle of rank r over X. Then it is well known (cf. e.g. [Fulton(1) ex. 14.6.6]) that we have for the Chow ring

$$A^*(Grass(m, E)) = \frac{A^*(X)[d_1, \ldots, d_{r-m}, e_1, \ldots, e_m]}{\left(\sum_{j=0}^{i} d_j e_{i-j} = c_i(E), \quad (1 \le i \le r) \right)},$$

where we have formally put $d_0 = 1, e_0 = 1$ and $d_j = 0, e_l = 0$ for $j > r - m, l > m$ respectively. One can summarize these relations to

$$(1 + d_1 + \ldots + d_{r-m})(1 + e_1 + \ldots + e_m) = c(E).$$

One has to note that the relation holds for every weight. We have

$$c(T_{r-m,E}) = (1 + d_1 + \ldots + d_{r-m}),$$
$$c(Q_{m,E}) = (1 + e_1 + \ldots + e_m).$$

In the case of a projective bundle $\mathbf{P}(E)$ we get in particular

$$A^*(\mathbf{P}(E)) = \frac{A^*(X)[P]}{\left(\sum_{i=0}^{r} c_i(E) P^{r-i} \right)},$$

where $P = c_1(\mathcal{O}_{\mathbf{P}(E)}(1))$.

For the Chern classes of a symmetric power of a vector bundle we have the well-known relation:

Remark 3.1.9. Let E be a vector bundle of rank r over X with total Chern class $c(E) = 1 + e_1 + \ldots e_r$. Let $c(E) = (1 + y_1) \ldots (1 + y_r)$ be a formal splitting of $c(E)$. Then we have

$$c(\text{Sym}^m(E)) = \prod_{i_1 \leq \ldots \leq i_m} (1 + y_{i_1} + \ldots + y_{i_m}).$$

If E has rank 2, we have

$$c(\text{Sym}^2(E)) = (1 + 2e_1 + 4e_2)(1 + e_1)$$
$$= 1 + 3e_1 + (2e_1^2 + 4e_2) + 4e_1 e_2$$
$$c(\text{Sym}^3(E)) = (1 + 3e_1 + 9e_2)(1 + 3e_1 + 2e_1^2 + e_2)$$
$$= 1 + 6e_1 + (11e_1^2 + 10e_2) + (6e_1^3 + 30e_1 e_2) + 18e_1^2 e_2 + 9e_2^2$$
$$c(\text{Sym}^4(E)) = (1 + 2e_1)(1 + 4e_1 + 16e_2)(1 + 4e_1 + 3e_1^2 + 4e_2)$$
$$= 1 + 10e_1 + (35e_1^2 + 20e_2) + (50e_1^3 + 120e_1 e_2)$$
$$+ (24e_1^4 + 208e_1^2 e_2 + 64e_2^2) + 96e_1^3 e_2 + 128e_1 e_2^2$$
$$c(\text{Sym}^5(E)) = (1 + 5e_1 + 25e_2)(1 + 5e_1 + 4e_1^2 + 9e_2)(1 + 5e_1 + 6e_1^2 + e_2)$$
$$= 1 + 15e_1 + (85e_1^2 + 35e_2) + (225e_1^3 + 350e_1 e_2)$$
$$+ (274e_1^4 + 1183e_1^2 e_2 + 259e_2^2)$$
$$+ (274e_1^5 + 1540e_1^3 e_2 + 1295e_1 e_2^2) + 600e_1^4 e_2 + 1450e_1^2 e_2^2 + 225e_2^3.$$

If E has rank 3, we get

$$c(\text{Sym}^2(E)) = (1 + 2e_1 + 4e_2 + 8e_3)(1 + 2e_1 + e_1^2 + e_1 e_2 - e_3)$$
$$= 1 + 4e_1 + (5e_1^2 + 5e_2) + (2e_1^3 + 11e_1 e_2 + 7e_3)$$
$$+ (6e_1^2 e_2 + 4e_2^2 + 14e_1 e_3) + (8e_1^2 e_3 + 4e_1 e_2^2 + 5e_2 e_3)$$
$$+ (8e_1 e_2 e_3 - 8e_3^2).$$

Definition 3.1.10. Let $y_1, \ldots y_r$ be variables and $f_1, \ldots, f_r$ the elementary symmetric polynomials in the y_i. Let

$$C^m(f_1, \ldots, f_r) := \prod_{i_1 \leq \ldots \leq i_m} (1 + y_{i_1} + \ldots y_{i_m})$$

viewed as a polynomial in the f_i. Each f_i has weight i. Let $C_i^m(f_1, \ldots, f_r)$ be the part of weight i in $C^m(f_1, \ldots, f_r)$.

From the above we see that for a vector bundle of rank r over X with Chern classes $e_1, \ldots, e_r$ the formula

$$c(\text{Sym}^m(E)) = C^m(e_1, \ldots, e_r)$$

holds. In future we don't want to distinguish between classes $a \in A^*(X)$ and $\pi_1^*(a) \in A^*(D_m^1(X))$ and also not between $b \in A^*(D_m^1(X))$ and $\pi_2^*(b) \in A^*(D_m^2(X))$.

Proposition 3.1.11.

$$A^*(D_m^2(X))$$

$$= \frac{A^*(X)\Big[d_1,\ldots,d_{d-m},e_1,\ldots,e_m,f_1,\ldots f_{d-m},g_1,\ldots,g_{\binom{m+1}{2}}\Big]}{\left(\begin{array}{c}(1+d_1+\ldots+d_{d-m})(1+e_1+\ldots+e_m)=\displaystyle\sum_{i=0}^{d}(-1)^i c_i(X),\\[2mm](1+f_1+\ldots+f_{d-m})(1+g_1+\ldots+g_{\binom{m+1}{2}})\\[2mm]=(1+d_1+\ldots+d_{d-m})C^2(e_1,\ldots,e_m)\end{array}\right)},$$

where

$$c(T_1)=(1+d_1+\ldots+d_{d-m}),$$
$$c(Q_1)=(1+e_1+\ldots+e_m),$$
$$c(T_2)=(1+f_1+\ldots+f_{d-m}),$$
$$c(Q_2)=(1+g_1+\ldots+g_{\binom{m+1}{2}}).$$

If X is a smooth projective variety over $\mathbf{C}$, the same result holds, if we replace the Chow ring $A^*(\cdot)$ by the cohomology ring $H^*(\cdot,\mathbb{Z})$ everywhere.

Proof: By the above $D_m^2(X)$ is isomorphic to the Grassmannian bundle $Grass((\binom{m+1}{2}),W_m^2(X))$ over $Grass(m,T_X^*)$. The exact sequence

$$0 \longrightarrow \operatorname{Sym}^2(Q_1) \longrightarrow W_m^2(X) \longrightarrow T_1 \longrightarrow 0$$

gives $c(W_m(X))=c(\operatorname{Sym}^2 Q_1)c(T_1)$. The result follows.   □

Two cases are somewhat simpler:

(1) the variety $D_{d-1}^2(X)$ of second order data of hypersurfaces on X.

(2) the variety $D_1^2(X)$ of second order data of curves on X.

Corollary 3.1.12. *For a variable P we write*

$$q_i(P) := \sum_{j \leq i}(-1)^j c_j(X)P^{i-j}, \quad 0 \leq i \leq d-1.$$

Then we have

$$A^*(D_{d-1}^2(X)) = \frac{A^*(X)[P,Q]}{\left(\begin{array}{c}\displaystyle\sum_{i=0}^{d}(-1)^i P^{d-i}c_i(X),\\[2mm](Q-P)\displaystyle\sum_{i=0}^{\binom{d}{2}}Q^{\binom{d}{2}-i}C_i^2(q_1(P),\ldots,q_{d-1}(P))\end{array}\right)},$$

where $P = c_1(\mathcal{O}_{\mathbf{P}(T_X^*)}(1))$, $Q = c_1(\mathcal{O}_{\mathbf{P}(W_{d-1}^2(X))}(1))$.

Proof: $D_{d-1}^2(X)$ is the projective bundle $\mathbf{P}(W_{d-1}^2(X))$ over $\mathbf{P}(T_X^*)$, and we have $c_i(Q_{d-1,T_X^*}) = q_i(P)$. Thus the result follows immediately from proposition 3.1.11.
□

Corollary 3.1.13.

$$A^*(D_1^2(X)) = \cfrac{H^*(X)[P,Q]}{\left(\begin{array}{l} \displaystyle\sum_{i=0}^{d} P^{d-i} c_i(X), \\[2mm] \displaystyle\sum_{i=0}^{d-1} \left(c_i(X) - \sum_{j \leq i-1} c_j(X) P^{i-j} \right) Q^{d-i} + 2c_d(X) \end{array} \right)},$$

where $P = c_1(\mathcal{O}_{\mathbf{P}(T_X)}(1))$, $Q = c_1(\mathcal{O}_{\mathbf{P}(W_1^2(X)^*)}(1))$.

Proof: This follows immediately from proposition 3.1.11.   □

If X is a smooth projective variety over $\mathbf{C}$, then corollaries 3.1.12 and 3.1.13 also hold, if we replace the Chow ring by the cohomology ring.

We will write the above formulas explicitely for X of dimension smaller or equal to four.

(1) Let $X = S$ be a smooth surface. Then we have

$$A^*(D_1^2(S)) = \cfrac{A^*(S)[P,Q]}{\left(\begin{array}{l} P^2 + c_1(S)P + c_2(S), \\ Q^2 + (c_1(S) - P)Q + 2c_2(S) \end{array} \right)}$$

where $P = c_1(\mathcal{O}_{\mathbf{P}(T_S)}(1))$, $Q = c_1(\mathcal{O}_{\mathbf{P}(W_1^2(S)^*)}(1))$.

(2) Let X be a smooth variety of dimension 3. Then we have

$$A^*(D_1^2(X)) = \cfrac{A^*(X)[P,Q]}{\left(\begin{array}{l} P^3 + c_1(X)P^2 + c_2(X)P + c_3(X), \\ Q^3 + (c_1(X) - P)Q^2 + (c_2(X) - c_1(X)P - P^2)Q + 2c_3(X) \end{array} \right)},$$

where $P = c_1(\mathcal{O}_{\mathbf{P}(T_X)}(1))$, $Q = c_1(\mathcal{O}_{\mathbf{P}(W_1^2(X)^*)}(1))$.

$$A^*(D_2^2(X) = \cfrac{A^*(X)[P,Q]}{\left(\begin{array}{l} P^3 - c_1(X)P^2 + c_2(X)P - c_3(X), \\ (Q-P)(Q+P-c_1(X))(Q^2 + 2(P-c_1(X))Q \\ \qquad + 4(P^2 - c_1(X)P + c_2(X)) \end{array} \right)},$$

where $P = c_1(\mathcal{O}_{\mathbf{P}(T_X^*)}(1))$, $Q = c_1(\mathcal{O}_{\mathbf{P}(W_2^2(X))}(1))$.

(3) Let X be a smooth variety of dimension 4. Then we have

$$A^*(D_1^2(X)) = \cfrac{A^*(X)[P,Q]}{\left(\begin{array}{l} P^4 + c_1(X)P^3 + c_2(X)P^2 + c_3(X)P + c_4(X), \\ Q^4 + (c_1(X) - P)Q^3 + (c_2(X) - c_1(X)P - P^2)Q^2 \\ \qquad + (c_3(X) - c_2(X)P - c_1(X)P^2 - P^3) + 2c_4(X) \end{array}\right)},$$

where $P = c_1(\mathcal{O}_{\mathbf{P}(T_X)}(1))$, $Q = c_1(\mathcal{O}_{\mathbf{P}(W_1^2(X)^*)}(1))$.

$$A^*(D_2^2(X)) = \frac{A^*(X)[p_1, p_2, r_1, r_2]}{(R_1, R_2, R_3, R_4)},$$

where

$$R_1 := p_1^3 - 2p_1 p_2 + p_1^2 c_1(X) - p_2 c_1(X) + p_1 c_2(X) + c_3(X),$$

$$R_2 := p_1^2 p_2 - p_2^2 + p_1 p_2 c_1(X) + p_2 c_2(X) - c_4(X),$$

$$R_3 := r_1^4 - 3r_1^2 r_2 + r_2^2 + r_1^3(-2p_1 + c_1(X)) + r_1 r_2(4p_1 - 2c_1(X))$$
$$\qquad + r_1^2(-2p_1 c_1(X) + 3p_2 + c_2(X)) + r_2(-3p_2 + 2p_1 c_1(X) - c_2(X))$$
$$\qquad + r_1(p_1 p_2 - 2p_1 c_2(X) + 3p_2 c_1(X) + c_3(X))$$
$$\qquad - 2p_2^2 + 2p_2 c_2(X) - 2p_1 c_3(X) + 2c_4(X),$$

$$R_4 := r_1^3 r_2 - 2r_1 r_2^2 + r_1^2 r_2(-2p_1 + c_1(X)) + r_2^2(2p_1 - c_1(X))$$
$$\qquad + r_1 r_2(3p_2 - 2p_1 c_1(X) + c_2(X))$$
$$\qquad + r_2(p_1 p_2 - 2p_1 c_2(X) + 3p_2 c_1(X) + c_3(X)) + 4p_1 c_4(X),$$

with

$$c(T_1) = (1 + p_1 + p_2),$$
$$c(T_2) = (1 + r_1 + r_2).$$

$A^*(D_3^2(X))$

$$= \cfrac{A^*(X)[P,Q]}{\left(\begin{array}{l} P^4 - c_1(X)P^3 + c_2(X)P^2 - c_3(X)P + c_4(X), \\ (Q - P)(Q^3 - 2(P - c_1(X))Q^2 + 4(P^2 - c_1(X)P + c_2(X))Q \\ \qquad + 8(P^3 - c_1(X)P^2 + c_2(X)P - c_3(X)) \\ \qquad \cdot (Q^3 + 2(P - c_1(X))Q^2 + (P^2 - 2c_1(X)P + c_1(X)^2)Q \\ \qquad - c_1(X)P^2 + c_1^2(X)P - c_1(X)c_2(X) - c_3(X)) \end{array}\right)},$$

where $P = c_1(\mathcal{O}_{\mathbf{P}(T_X^*)}(1))$, $Q = c_1(\mathcal{O}_{\mathbf{P}(W_1^2(X))}(1))$.

$D_m^2(X)$ **as the variety of second order data of m-dimensional subvarieties of X.**

We want to see in what respect $D_m^2(X)$ parametrizes the second order data of m-dimensional subvarieties of X. First we will more generally consider the lth order data of germs of smooth subvarieties.

Definition 3.1.14. Let Y be the germ of a smooth subvariety of dimension m at $x \in X$. Let $I_Y \subset \mathcal{O}_{X,x}$ be the ideal of Y in $\mathcal{O}_{X,x}$. The lth order datum of Y at x is the subscheme

$$D_{l,x}(Y) := spec(\mathcal{O}_{X,x}/(I_Y + \mathbf{m}_{X,x}^{l+1})).$$

Remark 3.1.15. The lth order data of germs of smooth subvarieties of X are the points of

$$Z_{\left(1,m,\binom{m+1}{2},\ldots,\binom{m+l-1}{l}\right)}(X) \subset X^{\left[\binom{m+l}{l}\right]}$$

(see 2.1.5, 2.1.6, 2.1.7).

Proof: For $Y \subset X$ a smooth subvariety defined in a neighbourhood of $x \in X$ we have

$$D_{l,x}(Y) \in Z_{\left(1,m,\binom{m+1}{2},\ldots,\binom{m+l-1}{l}\right)}(X).$$

Now let $Z \in Z_{\left(1,m,\ldots,\binom{m+l-1}{l}\right)}(X)$ and $supp(Z) = x \in X$. Let I_Z be the ideal of Z in $\mathcal{O}_{X,x}$. Then there are local parameters $(x_1, \ldots, x_d)$ near x such that

$$I_Z = (x_{m+1}, \ldots, x_d) + \mathbf{m}_{X,x}^{l+1}.$$

Let $Y \subset X$ be the smooth subvariety defined in a neighbourhood of x by the ideal $I_Y := (x_{m+1}, \ldots, x_d)$. Then we have $D_{l,x}(Y) = \mathcal{O}_{X,x}/I_Z$. $\quad\square$

Because of remark 3.1.15 we write

$$D_m^l(X)_0 := Z_{\left(1,m,\ldots,\binom{m+l-1}{l}\right)}(X).$$

We see that $D_1^l(X)_0 = X_{(n+1),c}^{[n+1]}$ (see remark 2.1.8). So $D_m^l(X)_0$ parametrizes l^{th} order data of smooth m-dimensional subvarieties of X. It is easy to see that $D_m^1(X)_0 = Grass(m, T_X^*)$. For $l \geq 2$ and $d \geq 2$ however $D_m^l(X)$ is not compact.

Remark 3.1.16. Let

$$\rho_l : D_m^l(X)_0 \longrightarrow D_m^{l-1}(X)_0$$
$$D_{l,x}(Y) \longmapsto D_{l-1,x}(Y)$$

Then $D_m^l(X)_0$ is via ρ_l a locally trivial fibre bundle over $D_m^{l-1}(X)_0$ with fibre $\mathbf{A}^{(d-m)\binom{m+l-1}{l}}$. This is only a reformulation of remark 2.1.7.

Now a variety of lth order data should be a natural smooth compactification of $D_m^l(X)_0$. This is for instance the case for $D_m^2(X)$, as this is given in a canonical way as a subscheme of a product of Hilbert schemes, it is smooth, compact and contains $D_m^2(X)_0$ as a dense open subvariety. There is a morphism

$$\phi_2 : D_m^2(X) \longrightarrow D_m^1(X) = Grass(m, T_X^*),$$

extending ρ_2. The fibres of ϕ_2 are obtained by compactifying the fibres of ρ_2 to the Grassmannian $Grass\big(\binom{m+1}{2}, \binom{m+1}{2} + (d-m)\big)$.

Now we want to compute the class of the complement $D_m^2(X)_\infty := D_m^2(X) \setminus D_m^2(X)_0$. It parametrizes in a suitable sense the second order data of singular m-dimensional subvarieties of X. We will use a tool that will play a major role in the enumerative applications of higher order data in section 3.2, the Porteous formula. We will not quote the result in full generality but in the formulation in which we are going to use it.

Definition 3.1.17. Let X be a smooth variety and E and F vector bundles on X of ranks e and f respectively. Let $c(E), c(F) \in A^*(X)$ be their total Chern classes. We write

$$c(F - E) := c(F)/c(E)$$

and $c_j(F - E)$ for the part of $c(F - E)$ lying in $A^j(X)$. The total *Segre class* $s(E)$ of E is given by

$$s(E) := c(-E) = 1/c(E),$$

and the jth Segre class $s_j(E)$ of E is the part of $s(E)$ in $A^j(X)$.

Let $\sigma : E \longrightarrow F$ be a morphism of vector bundles on X. For all $x \in X$ let $\sigma(x)$ be the corresponding map on the fibres. Let $\mathcal{D}_k(\sigma) \subset X$ be the subscheme

$$\mathcal{D}_k(\sigma) := \Big\{ x \in X \,\Big|\, rk(\sigma(x)) \leq k \Big\}$$

with its natural scheme structure, i.e. with respect to local trivialisations of E and F it is defined by the vanishing of minors of the matrix representing σ. We call $\mathcal{D}_k(\sigma)$ the kth degeneracy locus of σ. Let $[\mathcal{D}_k(\sigma)] \in A^*(X)$ be the class of $\mathcal{D}_k(\sigma)$. We call $[\mathcal{D}_k(\sigma)]$ the kth degeneracy cycle of σ.

Theorem 3.1.18[Fulton (1) Thm. 14.4].

(1) *Each irreducible component of* $\mathcal{D}_k(\sigma)$ *has codimension at most* $r := (e-k)(f-k)$ *in* X.

(2) *If the codimension of* $\mathcal{D}_k(\sigma)$ *in* X *is* r, *then we have:*

$$[\mathcal{D}_k(\sigma)] = det((c_{f-k+i-j}(F-E))_{1 \leq i,j \leq e-k}).$$

We consider the morphism

$$\psi : \pi_2^*(\mathrm{Sym}^2(Q_1)) \longrightarrow Q_2$$

of vector bundles on $D_m^2(X)$ which is defined by the diagram

$$
\begin{array}{ccccccccc}
 & & & & 0 & & & & \\
 & & & & \downarrow & & & & \\
 & & & & T_2 & & & & \\
 & & & & \downarrow & & & & \\
0 & \longrightarrow & \pi_2^*(\mathrm{Sym}^2(Q_1)) & \longrightarrow & W_m^2(X) & \longrightarrow & T_1 & \longrightarrow & 0 \\
 & & & \searrow\psi & \downarrow & & & & \\
 & & & & Q_2 & & & & \\
 & & & & \downarrow & & & & \\
 & & & & 0. & & & &
\end{array}
$$

Then $D_m^2(X)_\infty$ is the degeneracy cycle

$$\mathcal{D}_{\binom{m+1}{2}-1}(\psi) := \left\{ v \in D_m^2(X) \,\middle|\, \psi(v) \text{ is not onto} \right\}.$$

The intersection of each fibre $\pi_2^{-1}(v)$ with $D_m^2(X)_\infty$ is a divisor in $\pi_2^{-1}(v)$. So we get by the Porteous formula:

$$
\begin{aligned}
[D_m^2(X)_\infty] &= c_1(Q_2) - \pi_2^*(c_1(\mathrm{Sym}^2(Q_1))) \\
&= c_1(Q_2) - (m+1)\pi_2^*(c_1(Q_1)).
\end{aligned}
$$

Remark 3.1.19. Let $Y \subset X$ be a smooth locally closed subvariety of dimension $m_0 \geq m$. Then $D_m^2(Y)$ is in a natural way a locally closed subvariety of $D_m^2(X)$. If Y has dimension m, then $D_m^2(Y)$ is isomorphic to Y in a natural way.

Definition 3.1.20. Let $Y_1, Y_2 \subset X$ be smooth locally closed subschemes of dimensions $d_1, d_2 \geq m$ and $x \in X$. We say Y_1, Y_2 have l^{th} order contact along an

m-dimensional subvariety at x, if $x \in Y_1 \cap Y_2$, and there is a germ of a smooth subvariety $Z \subset X$ at x satisfying $D_{l,x}(Z) \subset Y_1$ and $D_{l,x}(Z) \subset Y_2$.

We say Y_1, Y_2 have m-dimensional l^{th} order contact, if there is an $x \in X$ such that they have l^{th} order contact along an m-dimensional subvariety at x. If m is the minimum $min(d_1, d_2)$, we say in this case that Y_1 and Y_2 have l^{th} order contact (at x).

From the definitions we get immediately:

Remark 3.1.21. Y_1 and Y_2 have m-dimensional l^{th} order contact at x, if and only if $D^l_m(Y_1)_0$, and $D^l_m(Y_2)_0$ intersect as subvarieties of $D^l_m(X)_0$ in points lying over $x \in X$.

In case $d_1 = m \le d_2$, Y_1 and Y_2 have second order contact at x if and only if $D^2_m(Y_1)$ and $D^2_m(Y_2)$ intersect as subvarieties of $D^2_m(X)$ in points lying over $x \in X$. (In this case the intersection point automatically lies in $D^2_m(Y_1)_0 \cap D^2_m(Y_2)_0$, as Y_1 is smooth of dimension m, and $D^2_m(X)_0 \cap D^2_m(Y_2) = D^2_m(Y_2)_0$.)

3.2. Varieties of higher order data and applications

We now want to try to generalize the definition of the varieties of second order data to a definition of varieties of higher order data. We will however only have partial success. This means that we give a general definition of the variety $D_m^n(X)$ of nth order data of m-dimensional subvarieties of a smooth variety X, which however does not behave very well in general. The varieties of third order data of curves and hypersurfaces on a smooth variety X turn out to be projective bundles over the corresponding varieties of second order data. However the varieties of third order data of subvarieties which have both dimension and codimension greater or equal to two are not locally trivial fibre bundles over the corresponding varieties of second order data. Also, even if X is a surface, $D_1^4(X)$ is not a locally trivial fibre bundle over $D_1^3(X)$. At the end of this section we give some enumerative applications of our results.

As a straightforward generalisation of the definition of the varieties of second order data of m-dimensional subvarieties of a smooth variety X we get the following definition:

Definition 3.2.1. Let X be a smooth projective variety of dimension d over a field k. Let $n, m \in \mathbb{Z}_{\geq 0}$ with $1 \leq m \leq d$. Let $\mathcal{D}_m^n(X)$ be the contravariant functor from the category of noetherian k-schemes to the category of sets which for noetherian k-schemes S, T and a morphism $\phi : S \longrightarrow T$ is given by:

$$
\mathcal{D}_m^n(X)(T) = \left\{ (Z_0, \ldots, Z_n) \;\middle|\; \begin{array}{c} Z_i \subset X \times T \text{ closed subscheme} \\ \text{flat of degree } \binom{m+i}{i} \text{ over T } (i = 0, \ldots, n) \\ Z_0 \subset Z_1 \subset \ldots \subset Z_n \\ \text{and } Z_i \cdot Z_j \supset Z_{i+j+1} \\ \text{for all } i, j \text{ with } i + j \leq n - 1 \end{array} \right\},
$$

$$
\begin{array}{cccc}
\mathcal{D}_m^n(X)(\phi) : & \mathcal{D}_m^n(X)(T) & \longrightarrow & \mathcal{D}_m^n(X)(S) \\
& (Z_0, \ldots, Z_n) & \longmapsto & (Z_0 \times_T S, \ldots, Z_n \times_T S).
\end{array}
$$

Here we use again the notations we have introduced in definition 3.1.1. In the same way as in lemma 3.1.2 we can show that $\mathcal{D}_m^n(X)$ is represented by a closed subscheme

$$
D_m^n(X) \subset X \times X^{[m+1]} \times \ldots \times X^{\left[\binom{m+n}{n}\right]}.
$$

We call $D_m^n(X)$ the variety of nth order data of m-dimensional subvarieties of X. $D_m^n(X)$ is as a subset of $X^{[m+1]} \times \ldots \times X^{\left[\binom{m+n}{n}\right]}$ given by:

$$
D_m^n(X) := \left\{ (Z_0, \ldots, Z_n) \in X \times X^{[m+1]} \times \ldots \times X^{\left[\binom{m+n}{n}\right]} \;\middle|\; \begin{array}{c} Z_0 \subset Z_1 \subset Z_2 \subset \ldots \subset Z_n \\ \text{and } Z_i \cdot Z_j \supset Z_{i+j+1} \\ \text{for all } i, j \text{ with } i + j \leq n - 1 \end{array} \right\}.
$$

Obviously we have $D_m^0(X) = X$. For $i = 0, \ldots, n$ let

$$r_i : D_m^n(X) \longrightarrow X^{\left[\binom{m+i}{i}\right]}$$

be the projection. We also consider the projection

$$\pi_n : D_m^n(X) \longrightarrow D_m^{n-1}(X).$$

It is not clear in which cases $D_m^n(X)$ is reduced, irreduble or smooth. In cases in which it is reducible a better candidat for the variety of higher order data is the closure the image of $D_m^n(X)_0$ under the obvious embedding. Let $\Delta \subset X \times X$ be the diagonal. Then $\text{Hilb}^{\binom{i+m}{i}}(\Delta^{i+1}/X)$ is a closed subscheme of $\text{Hilb}^{\binom{i+m}{i}}((X \times X)/X) = X^{\left[\binom{i+m}{i}\right]}$ for all i. We can see immediately that the projection $r_i : D_m^n(X) \longrightarrow X^{\left[\binom{m+i}{i}\right]}$ factors through $\text{Hilb}^{\binom{i+m}{i}}(\Delta^{i+1}/X)$, as we see from the definitions that $\text{Hilb}^{\binom{i+m}{i}}(\Delta^{i+1}/X)$ represents the functor

$$T \longmapsto \left\{ (Z_0, Z_i) \;\middle|\; \begin{array}{c} Z_0, Z_i \subset X \times T \text{ closed subschemes} \\ \text{flat of degrees 1, } \binom{m+i}{i} \text{ over } T \\ Z_i \subset Z_0^{i+1} \end{array} \right\}.$$

We now want to show that $D_1^3(X)$ and $D_{d-1}^3(X)$ are again Grassmannian bundles corresponding to vector bundles over $D_1^2(X)$ and $D_{d-1}^2(X)$ respectively. Before doing this we want to show that these two cases are the only ones in which we can expect such a result (exept for the trivial case $m = d$).

Remark 3.2.2.

(1) Let $2 \leq m \leq d - 2$. Then $\pi_3 : D_m^3(X) \longrightarrow D_m^2(X)$ is not a locally trivial fibre bundle.

(2) Let S be a smooth surface. Then $\pi_4 : D_1^4(S) \longrightarrow D_1^3(S)$ is not a locally trivial fibre bundle.

Proof: (1) Let $x \in X$. Let $x_1, \ldots x_d$ be local parameters near x and let

$$\mathbf{m}_{X,x} := (x_1, \ldots, x_d)$$

be the maximal ideal at x. Let Z_1, Z_2 be the subschemes of X with support x defined by the ideals

$$I_1 = (x_{m+1}, \ldots, x_d) + \mathbf{m}_{X,x}^2,$$
$$I_2 = (x_{m+1}, \ldots, x_d) + \mathbf{m}_{X,x}^3$$

in $\mathcal{O}_{X,x}$. Then we have $(x, Z_1, Z_2) \in D_m^2(X)$. The fibre $\pi_3^{-1}((x, Z_1, Z_2))$ consists exactly of the subschemes $Z_3 \subset X$ with support x whose ideal I_3 in $\mathcal{O}_{X,x}$ is of the form

$$I_3 = (V) + (x_{m+1}, \ldots, x_d) \cdot \mathbf{m}_{X,x} + \mathbf{m}_{X,x}^4$$

for some $(d-m)$-dimensional linear subspace V of

$$\langle x_{m+1} \ldots x_d \rangle + \langle x_i x_j x_l \mid i,j,l \le m \rangle.$$

(Here we denote by $\langle f_1, \ldots, f_r \rangle$ the span as a vector space in contrast to $(f_1, \ldots, f_r)$, which denotes the ideal generated by the f_i.) So we have

$$\pi_3^{-1}((x, Z_1, Z_2)) \cong Grass\left(\binom{m+2}{3}, d - m + \binom{m+2}{3}\right).$$

Let $Z_2' \subset X$ be the subscheme with support x defined by the ideal

$$I_2' := (x_{m+3}, \ldots, x_d, x_1^2, x_1 x_2) + (x_i x_j \mid i > m) + \mathbf{m}_{X,x}^3$$

in $\mathcal{O}_{X,x}$. Then (x, Z_1, Z_2') is a point of $D_m^2(X)$. The fibre $\pi_3^{-1}((x, Z_1, Z_2'))$ consists of exactly those subschemes Z_3' with support x whose ideal I_3' in $\mathcal{O}_{X,x}$ is of the form

$$\begin{aligned} I_3' =&(W) + (x_i x_j \mid i \ge m+3) + (x_i x_j \mid i,j \ge m+1) + (x_1^2 x_i, x_1 x_2 x_i \mid i \le d) \\ &+ (x_i x_j x_k \mid i \ge m+1) + \mathbf{m}_{X,x}^4 \end{aligned}$$

for an $\binom{m+2}{3}$-codimensional linear subspace W of

$$\begin{aligned} U :=& \langle x_{m+3}, \ldots, x_d, x_1^2, x_1 x_2 \rangle + \langle x_{m+1} x_i, x_{m+2} x_i \mid i \le m \rangle \\ &+ \langle x_i x_j x_l \mid 1 \le i < j \le l \le m; j > 2 \rangle. \end{aligned}$$

By $dim(U) = d - m + \binom{m+2}{3} + 1$ we have

$$\pi_3^{-1}((x, Z_1, Z_2')) \cong Grass\left(\binom{m+2}{3}, d - m + 1 + \binom{m+2}{3}\right).$$

(2) Now let S be a smooth surface, $s \in S$ and x, y local parameters near s. Let Z_1, Z_2, Z_3 be the subschemes of S with support s defined by

$$I_1 := (x, y^2),$$
$$I_2 := (x, y^3),$$
$$I_3 := (x, y^4).$$

Then we have $(s, Z_1, Z_2, Z_3) \in D_1^3(X)$. Thus $\pi_4^{-1}((s, Z_1, Z_2, Z_3))$ consists of the subschemes Z_4 with support s whose ideal I_4 in $\mathcal{O}_{S,s}$ is of the form $I_4 = w + (x^2, xy, y^5)$ for a one-dimensional linear subspace $w \subset \langle x, y^4 \rangle$. So we have

$$\pi_4^{-1}((s, Z_1, Z_2, Z_3)) \cong \mathbf{P}_1.$$

Let Z_2', Z_3' be the subschemes of S with support s defined by

$$I_2' := (x^2, xy, y^2),$$
$$I_3' := (x^2, xy, y^3).$$

Then (s, Z_1, Z_2', Z_3') is a point of $D_1^3(S)$. $\pi_4^{-1}((s, Z_1, Z_2', Z_3'))$ consists of the subschemes Z_4' with support s whose ideal is of the form

$$I_4' = (t) + (x^3, x^2y, xy^2, y^4)$$

for a two-dimensional linear subspace $t \subset \langle x^2, xy, y^3 \rangle$. So we have

$$\pi_4^{-1}((w, Z_1, Z_2, Z_3)) \cong \mathbf{P}_2. \qquad \square$$

Definition 3.2.3. Let X be a smooth projective variety of dimension d over a field k. Let m be a positive integer with $m \leq d$. We will again use the notations from the definitions 3.1.3, 3.1.4 and 3.1.5. Let $\overline{\pi}_2 := \tilde{\pi}_1 \circ \tilde{\pi}_2$. We define the subbundle $\overline{T}_2$ of $\overline{\pi}_2^*(J_3(X))$ by the diagram

$$
\begin{array}{ccccccccc}
 & & & & 0 & & 0 & & \\
 & & & & \uparrow & & \uparrow & & \\
0 & \longrightarrow & & & Q_2 & = & Q_2 & \longrightarrow & 0 \\
 & & & & \uparrow & & \uparrow & & \\
0 & \longrightarrow & \overline{\pi}_2^*(\mathrm{Sym}^3(T_X^*)) & \longrightarrow & \overline{\pi}_2^*(J_3(X)) & \longrightarrow & \overline{\pi}_2^*(J_2(X)) & \longrightarrow & 0 \\
 & & \| & & \uparrow & \square & \uparrow & & \\
0 & \longrightarrow & \overline{\pi}_2^*(\mathrm{Sym}^3(T_X^*)) & \longrightarrow & \overline{T}_2 & \longrightarrow & \tilde{T}_2 & \longrightarrow & 0 \\
 & & \uparrow & & \uparrow & & \uparrow & & \\
 & & 0 & & 0 & & 0 & &
\end{array}
$$

Let again $\Delta \subset X \times X$ be the diagonal and $\mathcal{I}_\Delta \subset \mathcal{O}_{X \times X}$ its ideal sheaf. Let

$$s_1, s_2 : X \times X \longrightarrow X$$

be the projections. For all non-negative integers $i \leq j$ let

$$J_j^i(X) := (s_2)_*((\mathcal{I}_\Delta)^i/(\mathcal{I}_\Delta)^{j+1}).$$

Then $J_j^i(X)$ is locally free, and we have the exact sequence

$$0 \longrightarrow J_j^i(X) \longrightarrow J_j(X) \longrightarrow J_{i-1}(X) \longrightarrow 0.$$

We see $J_j^0(X) = J_j(X)$ and $J_j^j(X) = \text{Sym}^j(T_X^*)$. Let $i_1 \leq i_2 \leq j_2 \leq j_1$ be positive integers. The multiplication in $\mathcal{O}_{X \times X}$ gives a morphism

$$\cdot : (\mathcal{I}_\Delta)^{i_1}/(\mathcal{I}_\Delta)^{j_1} \otimes (\mathcal{I}_\Delta)^{i_2}/(\mathcal{I}_\Delta)^{j_2} \longrightarrow (\mathcal{I}_\Delta)^{i_1+i_2}/(\mathcal{I}_\Delta)^{i_1+j_2}$$

of sheaves on $X \times X$. So it gives a morphism of locally free sheaves

$$\cdot : J_{j_1}^{i_1}(X) \otimes J_{j_2}^{i_2}(X) \longrightarrow J_{i_1+j_2}^{i_1+i_2}(X).$$

For locally free subsheaves $F \subset J_{j_1}^{i_1}(X)$, $G \subset J_{j_2}^{i_2}(X)$ we denote by $F \cdot G$ the image of $F \otimes G$ under "$\cdot$". This is a coherent subsheaf of $J_{i_1+j_2}^{i_1+i_2}(X)$. By definitions 3.1.4 and 3.1.5 we have

$$W_m^2(X) = \widetilde{T}_1/(T_1 \cdot \tilde{\pi}_1^*(T_X^*)),$$

and T_2 is a subbundle of $\tilde{\pi}_2^*(W_m^2(X))$. $\overline{T}_2$ is the preimage of

$$T_2 \subset \tilde{\pi}_2^*(W_m^2(X)) \subset \overline{\pi}_2^*(J_2(X))/(\tilde{\pi}_2^*(T_1) \cdot \overline{\pi}_2^*(T_X^*))$$

under the natural morphism

$$p : \overline{\pi}_2^*(J_3(X)) \longrightarrow \overline{\pi}_2^*(J_2(X))/(\tilde{\pi}_2^*(T_1) \cdot \overline{\pi}_2^*(T_X^*)) = \overline{\pi}_2^*(J_3(X))/(\tilde{\pi}_2^*(\widetilde{T}_1) \cdot \overline{\pi}_2^*(J_2^1(X))).$$

Here for coherent sheaves F, G we write $F \subset G$ to mean that F is a subsheaf of G. So $\tilde{\pi}_2^*(\widetilde{T}_1) \cdot \overline{\pi}_2^*(J_2^1(X))$ is a subbundle of $\overline{T}_2$, and we have

$$\overline{T}_2/(\tilde{\pi}_2^*(\widetilde{T}_1) \cdot \overline{\pi}_2^*(J_2^1(X))) = T_2.$$

From $\widetilde{T}_2 \subset \tilde{\pi}_2^*(\widetilde{T}_1) \subset \overline{\pi}_2^*(J_2^1(X))$ we get

$$\widetilde{T}_2 \cdot \overline{\pi}_2^*(J_2^1(X)) \subset \tilde{\pi}_2^*(\widetilde{T}_1) \cdot \overline{\pi}_2^*(J_2^1(X)),$$
$$\tilde{\pi}_2^*(\widetilde{T}_1) \cdot \tilde{\pi}_2^*(\widetilde{T}_1) \subset \tilde{\pi}_2^*(\widetilde{T}_1) \cdot \overline{\pi}_2^*(J_2^1(X)) \subset \overline{T}_2 \subset \overline{\pi}_2^*(J_3^1(X)).$$

We define the coherent sheaves $U_m^3, V_m^3, W_m^3(X)$ on $\widetilde{D}_m^2(X)$ by

$$U_m^3 := (\tilde{\pi}_2^*(\widetilde{T}_1) \cdot \overline{\pi}_2^*(J_2^1(X)))/(\tilde{\pi}_2^*(\widetilde{T}_1) \cdot \tilde{\pi}_2^*(\widetilde{T}_1)),$$
$$V_m^3 := (\tilde{\pi}_2^*(\widetilde{T}_1) \cdot \overline{\pi}_2^*(J_2^1(X)))/(\tilde{\pi}_2^*(\widetilde{T}_1) \cdot \tilde{\pi}_2^*(\widetilde{T}_1) + \widetilde{T}_2 \cdot \overline{\pi}_2^*(J_2^1(X))),$$
$$W_m^3(X) := (\overline{T}_2)/(\tilde{\pi}_2^*(\widetilde{T}_1) \cdot \tilde{\pi}_2^*(\widetilde{T}_1) + (\widetilde{T}_2) \cdot \overline{\pi}_2^*(J_2^1(X))).$$

Then we obviously have the exact sequence

$$0 \longrightarrow V_m^3 \longrightarrow W_m^3(X) \longrightarrow T_2 \longrightarrow 0.$$

Lemma 3.2.4. *Let $m = 1$ or $m = d - 1$.*

(1) U_m^3 is locally free of rank $m(d-m) + \binom{m+2}{3}$.

(2) V_m^3 is locally free of rank $\binom{m+2}{3}$.

(3) $W_m^3(X)$ is locally free of rank $d - m + \binom{m+2}{3}$.

Proof: By the exact sequence

$$0 \longrightarrow V_m^3 \longrightarrow W_m^3(X) \longrightarrow T_2 \longrightarrow 0$$

it is enough to show these results for U_m^3 and V_m^3. It is enough to check them fibrewise. Let $v \in D_m^2(X)$ be a point lying over $x \in X$. Let $x_1, \ldots, x_d$ be local parameters near x. For $i = 1, \ldots, d$ we denote by $\bar{x}_i$ the class of x_i in $\mathcal{O}_{X,x}/\mathbf{m}_{X,x}^4$. We can assume that the fibre $\tilde{\pi}_2^*(T_1(v))$ is of the form $(x_{m+1}, \ldots, x_d)/\mathbf{m}_{X,x}^2$. Then we have for the fibres:

$$(\tilde{\pi}_2^* \widetilde{T}_1 \cdot \tilde{\pi}_2^* \tilde{\pi}_2^* J_2^1(X))(v) = \langle \bar{x}_i \bar{x}_j \,|\, i > m \rangle + \langle \bar{x}_i \bar{x}_j \bar{x}_l \,|\, i,j,l \leq d \rangle,$$
$$(\tilde{\pi}_2^*(\widetilde{T}_1) \cdot \tilde{\pi}_2^*(\widetilde{T}_1))(v) = \langle \bar{x}_i \bar{x}_j \,|\, i,j > m \rangle + \langle \bar{x}_i \bar{x}_j \bar{x}_l \,|\, i > m \rangle.$$

Let $A_0 := \langle \bar{x}_i \bar{x}_j \,|\, i \leq m, j > m \rangle + \langle \bar{x}_i \bar{x}_j \bar{x}_l \,|\, i,j,l \leq m \rangle$. Then the restriction of the natural projection

$$s : (\tilde{\pi}_2^* \widetilde{T}_1 \cdot \tilde{\pi}_2^* J_2^1(X))(v) \longrightarrow U_m^3(v)$$

to A_0 is an isomorphism, and (1) follows.

$T_2(v)$ is an $(m-d)$-dimensional linear subspace of

$$\tilde{\pi}_2^*(W_m^2(X))(v) = (\tilde{\pi}_2^* \widetilde{T}_1 / (\tilde{\pi}_2^* T_1 \cdot \tilde{\pi}_2^* T_X^*))(v).$$

Let $p : \tilde{\pi}_2^*(W_m^2(X))(v) \longrightarrow \tilde{\pi}_2^*(T_1)(v)$ be the projection. As we have assumed that $m = 1$ or $m = d - 1$ holds, we have either $p(T_2(v)) = \tilde{\pi}_2^*(T_1)(v)$, or $p(T_2(v))$ has codimension 1 in $\tilde{\pi}_2^*(T_1)(v)$. ($\tilde{\pi}_2^*(T_1)(v)$ is one-dimensional in case $m = d - 1$, and $T_2(v)$ has codimension 1 in $\tilde{\pi}_2^*(W_m^2(X))(v)$ in case $m = 1$.)

(a) p is onto. Then we have

$$\widetilde{T}_2(v) = (y_{m+1}, \ldots, y_d)/\mathbf{m}_{X,x}^3,$$

where $x_1, \ldots, x_m, y_{m+1}, \ldots, y_d$ are local parameters near x. So we can assume that $x_i = y_i$ for $i = m+1, \ldots, d$. Then we have

$$(\widetilde{T}_2 \cdot \tilde{\pi}_2^*(J_2^1(X)))(v) = \langle \bar{x}_i \bar{x}_j \,|\, i > m \rangle + \langle \bar{x}_i \bar{x}_j \bar{x}_l \,|\, i > m \rangle.$$

Let $A_1 := \langle \bar{x}_i \bar{x}_j \bar{x}_l \,|\, i,j,l \leq m \rangle$. Then the restriction of the natural projection

$$q_1 : (\tilde{\pi}_2^*(\widetilde{T}_1) \cdot \tilde{\pi}_2^*(J_2^1(X)))(v) \longrightarrow V_m^3(v)$$

to A_1 is an isomorphism, and (2) follows.

(b) $p(T_2(v))$ has codimension 1. By changing the local coordinates if neccessary we can assume

$$\widetilde{T}_2(v) = ((x_{m+2}, \ldots, x_d, f) + (x_{m+1}x_j \,|\, j \le m+1))/\mathbf{m}_{X,x}^3$$

for an $f \in (x_i x_j \,|\, i,j \le m) \setminus \mathbf{m}_{X,x}^3$. Let $\bar{f}$ denote the class of f in $\mathcal{O}_{X,x}/\mathbf{m}_{X,x}^4$. Then we have

$$(\widetilde{T}_2 \cdot \overline{\pi}_2^* J_2^1(X))(v) = \langle \bar{x}_i \bar{x}_j \,|\, j \ge m+2 \rangle + \langle \bar{x}_i \bar{x}_j \bar{x}_l \,|\, i > m \rangle + \langle \bar{f} \bar{x}_l \,|\, l \le m \rangle.$$

Let $A_2 := \langle \bar{x}_{m+1} \bar{x}_i \,|\, i \le m \rangle + \langle \bar{x}_i \bar{x}_j \bar{x}_l \,|\, i,j,l \le m \rangle$. Then the restriction of the natural projection

$$q_2 : (\tilde{\pi}_2^*(\widetilde{T}_1) \cdot \overline{\pi}_2^*(J_2^1(X)))(v) \longrightarrow V_m^3(v)$$

to A_1 is a surjection with kernel $\langle \bar{f} \bar{x}_i \,|\, i \le m \rangle$, and (2) follows. $\quad\square$

Definition 3.2.5. We put

$$\widetilde{D}_1^3(X) := \check{\mathbf{P}}(W_1^3(X)) = \mathbf{P}((W_1^3(X))^*),$$
$$\widetilde{D}_{d-1}^3(X) := \mathbf{P}(W_{d-1}^3(X)) = \check{\mathbf{P}}((W_{d-1}^3(X))^*).$$

For $m = 1$ or $m = d-1$ let $\tilde{\pi}_3 : \widetilde{D}_m^3(X) \longrightarrow \widetilde{D}_m^2(X)$ be the projection and $\overline{\pi}_3 := \tilde{\pi}_1 \circ \tilde{\pi}_2 \circ \tilde{\pi}_3$. Let $T_3 := T_{d-m, W_m^3(X)}$ be the tautological subbundle and

$$0 \longrightarrow T_3 \xrightarrow{i_3} \tilde{\pi}_3^*(W_m^3(X)) \xrightarrow{q_3} Q_3 \longrightarrow 0$$

the canonical exact sequence. Let

$$K := \tilde{\pi}_2^*(\widetilde{T}_1) \cdot \tilde{\pi}_2^*(\widetilde{T}_1) + \widetilde{T}_2 \cdot \overline{\pi}_2^*(J_2^1(X))$$

be the kernel of the natural vector bundle morphism $\overline{p} : \overline{T}_2 \longrightarrow W_m^3(X)$. We define the vector bundle $\widetilde{T}_3$ on $\widetilde{D}_m^3(X)$ by the diagram

$$
\begin{array}{ccccccccc}
 & & 0 & & 0 & & 0 & & \\
 & & \uparrow & & \uparrow & & \uparrow & & \\
0 & \longrightarrow & T_3 & \xrightarrow{i_3} & \tilde{\pi}_3^*(W_m^3(X)) & \xrightarrow{q_3} & Q_3 & \longrightarrow & 0 \\
 & & \uparrow & \square & \uparrow{\scriptstyle \tilde{\pi}_3^*(\overline{p})} & & \| & & \\
0 & \longrightarrow & \widetilde{T}_3 & \xrightarrow{\tilde{i}_3} & \tilde{\pi}_3^*(\overline{T}_2) & \xrightarrow{\tilde{q}_3} & Q_3 & \longrightarrow & 0 \\
 & & \uparrow & & \uparrow & & \uparrow & & \\
0 & \longrightarrow & \tilde{\pi}_3^*(K) & = & \tilde{\pi}_3^*(K) & \longrightarrow & 0 & & \\
 & & \uparrow & & \uparrow & & & & \\
 & & 0 & & 0 & & & &
\end{array}
$$

and the vector bundle $\widetilde{Q}_3$ on $\widetilde{D}_m^3(X)$ by

$$
\begin{array}{ccccccc}
 & & 0 & & 0 & & 0 \\
 & & \downarrow & & \downarrow & & \downarrow \\
0 & \longrightarrow & \widetilde{T}_3 & \xrightarrow{\tilde{i}_3} & \tilde{\pi}_3^*(\overline{T}_2) & \xrightarrow{\hat{q}_3} & Q_3 & \longrightarrow & 0 \\
 & & \| & & \downarrow{\scriptstyle j_3} & & \downarrow{\scriptstyle j_3} \\
0 & \longrightarrow & \widetilde{T}_3 & \longrightarrow & \tilde{\pi}_3^*(J_3(X)) & \longrightarrow & \widetilde{Q}_3 & \longrightarrow & 0 \\
 & & \downarrow & & \downarrow & & \downarrow{\scriptstyle \hat{q}_3} \\
0 & \longrightarrow & & & \tilde{\pi}_3^*(\widetilde{Q}_2) & = = & \tilde{\pi}_3^*(\widetilde{Q}_2) & \longrightarrow & 0 \\
 & & & & \downarrow & & \downarrow \\
 & & & & 0 & & 0
\end{array}
$$

In particular we get the exact sequence

$$
0 \longrightarrow Q_3 \xrightarrow{j_3} \widetilde{Q}_3 \xrightarrow{\hat{q}_3} \tilde{\pi}_3^*(\widetilde{Q}_2) \longrightarrow 0.
$$

We now generalize the definition of the bundles $(\mathcal{O}_X)_m^1$ and $(\mathcal{O}_X)_m^2$ from 3.1.3:

Definition 3.2.6. For any $l \in \mathbb{N}$ let again $Z_l(X) \subset X \times X^{[l]}$ be the universal subscheme with the projections

$$
\begin{array}{c}
Z_l(X) \\
\swarrow{\scriptstyle p_l} \qquad \searrow{\scriptstyle q_l} \\
X \qquad\qquad\qquad X^{[l]}.
\end{array}
$$

For any vector bundle E of rank r on X we put

$$
\widetilde{E}_l := (q_l)_*(p_l^*(E)).
$$

This is a vector bundle of rank rl on $X^{[l]}$. For all $n \in \mathbb{N}$ and all $m \le d$ we put

$$
(E)_m^n := r_n^*(\widetilde{E}_{\binom{m+n}{n}}),
$$

where $r_n : D_m^n(X) \longrightarrow X^{\left[\binom{m+n}{n}\right]}$ is the projection from 3.2.1. $(E)_m^n$ is a vector bundle of rank $r \cdot \binom{m+n}{n}$ on $D_m^n(X)$. We call it the contact bundle corresponding to E, X and m.

Theorem 3.2.7. *Let $m = 1$ or $m = d - 1$. Then there is an isomorphism*

$$\phi_3 := D_m^3(X) \longrightarrow \tilde{D}_m^3(X),$$

for which the diagram

$$
\begin{array}{ccc}
D_m^3(X) & \xrightarrow{\ \phi_3\ } & \tilde{D}_m^3(X) \\
\downarrow{\scriptstyle \pi_3} & & \downarrow{\scriptstyle \tilde{\pi}_3} \\
D_m^2(X) & \xrightarrow{\ \phi_2\ } & \tilde{D}_m^2(X)
\end{array}
$$

commutes such that $\phi^*(\tilde{Q}_3) = (\mathcal{O}_X)_m^3$.

Proof: We use the notations from definition 3.1.6. Then we can write $\mathcal{D}_m^3(X)$ as

$$
\mathcal{D}_m^3(X)(T) = \left\{ (\phi, Z_1, Z_2, Z_3) \;\middle|\;
\begin{array}{c}
\phi : T \longrightarrow X \text{ morphism over } k \\
Z_1, Z_2, Z_3 \subset X \times T \\
\text{closed subschemes flat of degrees} \\
m + 1, \binom{m+2}{2}, \binom{m+3}{3} \text{ over } T \text{ with} \\
\Delta_\phi \subset Z_1 \subset Z_2 \subset Z_3, \\
Z_1 \subset \Delta_\phi^2, \; Z_2 \subset \Delta_\phi \cdot Z_1, \\
Z_3 \subset \Delta_\phi \cdot Z_2, \; Z_3 \subset Z_1 \cdot Z_1.
\end{array}
\right\}.
$$

Let

$$Z_1, Z_2, Z_3 \subset X \times D_m^3(X)$$

be the universal families of degrees $m + 1$, $\binom{m+2}{2}$, $\binom{m+3}{3}$ over $D_m^3(X)$. Via ϕ_2 we identify $\tilde{D}_m^2(X)$ with $D_m^2(X)$ and $\tilde{\pi}_1$ and $\tilde{\pi}_2$ with π_1 and π_2 respectively. We put

$$\pi := \pi_1 \circ \pi_2 \circ \pi_3,$$

$$\tilde{\pi} := \tilde{\pi}_1 \circ \tilde{\pi}_2 \circ \tilde{\pi}_3 = \pi_1 \circ \pi_2 \circ \tilde{\pi}_3.$$

The subvariety $\Delta_\pi \subset X \times D_m^3(X)$ is via the projection p to the second factor isomorphic to $D_m^3(X)$, and we have

$$p_*(\mathcal{O}_{\Delta_\pi^{n+1}}) = \pi^*(J_n(X)).$$

For a subscheme $Z \subset X \times D_m^3(X)$ let $\mathcal{I}_Z$ be the ideal of Z in $X \times D_m^3(X)$. By definition we have

$$Z_2 \subset Z_3 \subset \Delta_\pi \cdot Z_2 \subset \Delta_\pi^4,$$

$$Z_3 \subset Z_1 \cdot Z_1 \subset \Delta_\pi^4.$$

So $p_*(\mathcal{I}_{Z_2/Z_3})$ is a locally free quotient of rank $\binom{m+2}{3}$ of

$$p_*(\mathcal{I}_{Z_2}/(\mathcal{I}_{\Delta_\pi} \cdot Z_2 + \mathcal{I}_{Z_1 \cdot Z_1})) = \pi_3^*(W_m^3(X)).$$

This defines a morphism $\phi_3 : D_m^3(X) \longrightarrow \tilde{D}_m^3(X)$ over $D_m^2(X)$.

Let $Z_1 := \tilde{\pi}_3^{-1}(W_1)$, $Z_2 := \tilde{\pi}_3^{-1}(W_2)$ for the universal subschemes W_1 and W_2 over $D_m^2(X)$ of degrees $m + 1$ and $\binom{m+2}{2}$ respectively. The subvariety $\Delta_{\tilde{\pi}} \subset X \times \tilde{D}_m^3(X)$ is via the projection $\tilde{p}$ to the second factor isomorphic to $\tilde{D}_m^3(X)$, and we have

$$\tilde{p}_*(\mathcal{O}_{\Delta_{\tilde{\pi}}^{n+1}}) = \tilde{\pi}^*(J_n(X)).$$

$\tilde{T}_3$ is an $\mathcal{O}_{\tilde{D}_m^3(X)}$-submodule of $\tilde{\pi}^*(J_4(X))$. Let $I_3 \subset \mathcal{O}_{\Delta_{\tilde{\pi}}^4}$ be the $\mathcal{O}_{\Delta_*}$-submodule with $\tilde{p}_*(I_3) = \tilde{T}_3$. By the inclusions

$$\tilde{\pi}^*(J_2^1(X)) \cdot \tilde{\pi}_3^*(\tilde{T}_2) \subset \tilde{T}_3,$$
$$\tilde{\pi}_3^*(\pi_2^*(\tilde{T}_1)) \cdot \tilde{\pi}_3^*(\pi_2^*(\tilde{T}_1)) \subset \tilde{T}_3,$$
$$\tilde{T}_3 \subset \tilde{\pi}_3^*(\overline{T}_2)$$

(see definition 3.2.5) we have

$$\mathcal{I}_{\Delta_*/\Delta_{\tilde{\pi}}^4} \cdot \mathcal{I}_{Z_2/\Delta_{\tilde{\pi}}^4} \subset I_3,$$
$$\mathcal{I}_{Z_1/\Delta_{\tilde{\pi}}^4}^2 \subset I_3,$$
$$I_3 \subset \mathcal{I}_{Z_2/\Delta_{\tilde{\pi}}^4}.$$

So we have in particular $\mathcal{O}_{\Delta_{\tilde{\pi}}^4} \cdot I_3 = I_3$. So I_3 is an ideal in $\mathcal{O}_{\Delta_{\tilde{\pi}}^4}$ and defines a subscheme $Z_3 \subset \Delta_{\tilde{\pi}}^4$ satisfying

$$\tilde{p}_*(\mathcal{O}_{Z_3}) = \tilde{\pi}^*(J_3(X))/\tilde{T}_3 = \tilde{Q}_3.$$

By the inclusions

$$\mathcal{I}_{\Delta_*/\Delta_{\tilde{\pi}}^4} \cdot \mathcal{I}_{Z_2/\Delta_{\tilde{\pi}}^4} \subset I_3,$$
$$\mathcal{I}_{Z_1/\Delta_{\tilde{\pi}}^4}^2 \subset I_3,$$
$$I_3 \subset \mathcal{I}_{Z_2/\Delta_{\tilde{\pi}}^4}$$

we have

$$Z_2 \subset Z_3 \subset \Delta_{\tilde{\pi}} \cdot Z_2,$$
$$Z_3 \subset Z_1 \cdot Z_1,$$
$$Z_2 \subset Z_3.$$

$(\tilde{\pi}, Z_1, Z_2, Z_3)$ defines a morphism $\psi_3 : \tilde{D}_m^3(X) \longrightarrow D_m^3(X)$ over $D_m^2(X)$ satisfying $\psi_3^*((\mathcal{O}_X)_m^3) = \tilde{Q}_3$. It is easy to see that $\psi_3 = \phi_3^{-1}$. $\quad\square$

In future we want to identify $D_1^3(X)$ with $\tilde{D}_1^3(X)$ and also $D_{d-1}^3(X)$ with $\tilde{D}_{d-1}^3(X)$.

As $D_1^3(X)$ is smooth, we see first that the projection $r_3 : D_1^3(X) \longrightarrow X^{[4]}$ factors through $X_{(4)}^{[4]}$. As also $X_{(4),c}^{[4]} = Z_{(1,1,1,1)}(X)$ is smooth, r_3 is an isomorphism over $Z_{(1,1,1,1)}(X)$. The preimage $D_1^3(X)_0$ parametrizes third order data of (germs of) smooth curves on X. Here the nth order datum of a smooth subvariety $Y \subset X$ in a point $x \in Y$ is the quotient $\mathcal{O}_{Y,x}/\mathfrak{m}_{X,x}^{n+1}$ of $\mathcal{O}_{X,x}$. In a similar way one can treat $D_{d-1}^3(X)$: the preimage $r_3^{-1}(Z_{\left(1,d-1,\binom{4}{2},\binom{4+1}{3}\right)}(X))$ is an open dense subset $D_{d-1}^3(X)_0$ in $D_{d-1}^3(X)$, and the restriction $r_3|_{D_{d-1}^3(X)_0}$ is an isomorphism. $r_3^{-1}(D_{d-1}^3(X)_0)$ parametrizes third order data of (germs of) smooth hypersurfaces of X.

Remark 3.2.8. Let $Y \subset X$ be a smooth closed subvariety, Then for all $n \in I\!N$ the Hilbert scheme $Y^{[n]}$ is a closed subscheme of $X^{[n]}$. So for all $n, m \in I\!N$ with $m \leq d$ $D_m^n(Y)$ is a closed subscheme of $D_m^n(X)$. From the definitions of the vector bundles $(\mathcal{O}_X)_m^n$ and $(\mathcal{O}_Y)_m^n$ we see that $(\mathcal{O}_X)_m^n|_{D_m^n(Y)} = (\mathcal{O}_Y)_m^n$. So $\tilde{D}_m^2(Y) \subset \tilde{D}_m^2(X)$ and $\tilde{D}_1^3(Y) \subset \tilde{D}_1^3(X)$ are closed subvarieties with

$$\tilde{Q}_i(X)|_Y = \tilde{Q}_i(Y), \quad Q_i(X)|_Y = Q_i(Y), \qquad (i = 1, 2, 3).$$

Here we write $Q_i(X)$, $\tilde{Q}_i(X)$ for the classes Q_i, $\tilde{Q}_i$ on $D_m^i(X)$ and similar for Y.

In case $m = d = dim(X)$ we see immediately that $D_m^n(X)$ is isomorphic to X via its projection. The universal families are $Z_i = \Delta^{i+1} \subset X \times X$ for $i = 0, \ldots, n$. So we have

$$(\mathcal{O}_X)_m^n = J_n(X),$$
$$(\mathcal{O}_X)_m^n/(\mathcal{O}_X)_m^{n+1} = \text{Sym}^n(T_X^*).$$

Now we can compute the Chow rings of $D_1^3(X)$ and $D_{d-1}^3(X)$. For this we first have to determine the Chern classes of $W_1^3(X)$ and $W_{d-1}^3(X)$.

Lemma 3.2.9.

(1) *In case $m = 1$ we have $V_1^3 \cong \tilde{\pi}_2^*(Q_1) \otimes Q_2$, and so there is an exact sequence*

$$0 \longrightarrow \tilde{\pi}_2^*(Q_1) \otimes Q_2 \longrightarrow W_1^3(X) \longrightarrow T_2 \longrightarrow 0$$

on $\tilde{D}_1^2(X)$.

(2) *In case $m = d - 1$ there are exact sequences*

$$0 \longrightarrow T_2 \otimes \tilde{\pi}_2^*(Q_1) \longrightarrow U_{d-1}^3 \longrightarrow V_{d-1}^3 \longrightarrow 0$$

$$0 \longrightarrow \tilde{\pi}_2^*(\text{Sym}^3(Q_1)) \longrightarrow U_{d-1}^3 \longrightarrow \tilde{\pi}_2^*(T_1) \otimes \tilde{\pi}_2^*(Q_1) \longrightarrow 0$$

$$0 \longrightarrow V_{d-1}^3 \longrightarrow W_{d-1}^3(X) \longrightarrow T_2 \longrightarrow 0$$

on $\tilde{D}_{d-1}^2(X)$.

Proof:

(1) Let $w : \tilde{\pi}_2^*(\widetilde{T}_1) \otimes \overline{\pi}_2^*(J_2^1(X)) \longrightarrow V_1^3$ be the natural homomorphism. We see immediately that w is onto and

$$\overline{\pi}_2^*(J_2^1(X)) \otimes \widetilde{T}_2 + \tilde{\pi}_2^*(\widetilde{T}_1) \otimes \tilde{\pi}_2^*(\widetilde{T}_1)$$

lies in the kernel of w. As all the sheaves we are considering are locally free and have the right rank, we have

$$ker(w) = \overline{\pi}_2^*(J_2^1(X)) \otimes \widetilde{T}_2 + \tilde{\pi}_2^*(\widetilde{T}_1) \otimes \tilde{\pi}_2^*(\widetilde{T}_1),$$

and obviously this is also the kernel of the natural map $\bar{w} : \tilde{\pi}_2^*(\widetilde{T}_1) \otimes \overline{\pi}_2^*(J_2^1(X)) \longrightarrow \tilde{\pi}_2^*(Q_1) \otimes Q_2$.

(2) We already know the lower sequence. The middle sequence comes from the diagram

$$
\begin{array}{ccccccccc}
 & & 0 & & 0 & & 0 & & \\
 & & \downarrow & & \downarrow & & \downarrow & & \\
0 & \to & \tilde{\pi}_2^*(T_1)\cdot\overline{\pi}_2^*(J_2^2(X)) & \to & \tilde{\pi}_2^*(\widetilde{T}_1)\cdot\tilde{\pi}_2^*(\widetilde{T}_1) & \to & \tilde{\pi}_2^*(\mathrm{Sym}^2(T_1)) & \to & 0 \\
 & & \downarrow & & \downarrow & & \downarrow & & \\
0 & \to & \overline{\pi}_2^*(\mathrm{Sym}^3(T_X^*)) & \to & \tilde{\pi}_2^*(\widetilde{T}_1)\cdot\overline{\pi}_2^*(J_2^1(X)) & \to & \tilde{\pi}_2^*(T_1)\cdot\overline{\pi}_2^*(T_X^*) & \to & 0 \\
 & & \downarrow & & \downarrow & & \downarrow & & \\
0 & \to & \tilde{\pi}_2^*(\mathrm{Sym}^3(Q_1)) & \to & U_{d-1}^3 & \to & (\tilde{\pi}_2^*(T_1)\cdot\overline{\pi}_2^*(T_X^*))/\mathrm{Sym}^2(\tilde{\pi}_2^*(T_1)) & \to & 0 \\
 & & \downarrow & & \downarrow & & \downarrow & & \\
 & & 0 & & 0 & & 0 & &
\end{array}
$$

if we use $(T_1 \cdot \tilde{\pi}_1^*(T_X^*))/\mathrm{Sym}^2(T_1) \cong T_1 \otimes Q_1$.

Let $w_2 : U_{d-1}^3 \longrightarrow V_{d-1}^3$ be the natural homomorphism. We have to show $ker(w_2) \cong T_2 \otimes \tilde{\pi}_2^*(Q_1)$. We consider the exact sequence

$$\widetilde{T}_2 \otimes \overline{\pi}_2^*(J_2^1(X)) \xrightarrow{w_0} U_{d-1}^3 \xrightarrow{w_2} V_{d-1}^3 \longrightarrow 0,$$

where w_0 is the obvious map. We see that

$$(\tilde{\pi}_2^*(T_1) \cdot \overline{\pi}_2^*(T_X^*)) \otimes \overline{\pi}_2^*(J_2^1(X)) + \widetilde{T}_2 \otimes \tilde{\pi}_2^*(\widetilde{T}_1) \subset ker(w_0)$$

and

$$(\widetilde{T}_2 \otimes \overline{\pi}_2^*(J_2^1(X)))/((\tilde{\pi}_2^*(T_1) \cdot \overline{\pi}_2^*(T_X^*)) \otimes \overline{\pi}_2^*(J_2^1(X)) + \widetilde{T}_2 \otimes \tilde{\pi}_2^*(\widetilde{T}_1))$$
$$\cong \widetilde{T}_2 \otimes \overline{\pi}_2^*(T_X)/((\tilde{\pi}_2^*(T_1) \cdot \overline{\pi}_2^*(T_X^*)) \otimes \overline{\pi}_2^*(T_X^*) + \widetilde{T}_2 \otimes \tilde{\pi}_2^*(T_1))$$
$$\cong T_2 \otimes \tilde{\pi}_2^*(Q_1).$$

So there is a surjection of vector bundles

$$w_1 : T_2 \otimes \tilde{\pi}_2^*(Q_1) \longrightarrow ker(w_2).$$

Because the bundles have the same rank, it is an isomorphism. So (2) follows. □

Again for $i = 1, 2, 3$ we don't want to distinguish notationally between a_0 in $A^*(D_m^{i-1}(X))$ and $\pi_i^*(a_0)$. We formulate our results (proposition 3.2.10 and proposition 3.2.11) only for the Chow rings, but it is clear that they also hold if we replace the Chow rings by the cohomology rings everywhere.

Proposition 3.2.10. *Let X be a smooth projective variety of dimension d. Then*

$$A^*(D_1^3(X)) = \frac{A^*(X)[P, Q, R]}{\left(\begin{array}{c} \displaystyle\sum_{i=0}^{d} P^{d-i} c_i(X), \\[2em] \displaystyle\sum_{i=0}^{d-1} \left(c_i(X) - \sum_{j \le i-1} c_j(X) P^{i-j} \right) Q^{d-i} + 2c_d(X) \\[2em] \displaystyle\sum_{i=0}^{d} \left(c_i(X) - 2c_{i-1}(X)P - \sum_{j=0}^{i-2} c_j(X) \right. \\[2em] \left. \cdot \left(PQ^{i-1-j} - \sum_{l=1}^{i-2-j} Q^l P^{i-j-l} \right) \right) R^{d-i} \end{array} \right)}$$

Here $P = c_1(\mathcal{O}_{\mathbf{P}(T_X)}(1))$, $Q = c_1(\mathcal{O}_{\mathbf{P}(W_1^2(X)^*)}(1))$, $R = c_1(\mathcal{O}_{\mathbf{P}(W_1^3(X)^*)}(1))$.

Proof: This follows immediately from lemma 3.2.9(1). □

Proposition 3.2.11. *Let X be a smooth projective variety of dimension d. As an abbreviation we write* $q_i(P) := \sum_{j \le i}(-1)^j c_j(X) P^{i-j}$, $0 \le i \le d-1$. *Then we have with the notations of definition 3.1.10 and corollary 3.1.12*

$$A^*(D_{d-1}^3(X)) = \frac{A^*(X)[P, Q, R]}{\left(\begin{array}{c} \displaystyle\sum_{i=0}^{d} (-1)^i P^{n-i} c_i(X), \\[2em] \displaystyle (Q-P) \sum_{i=0}^{\binom{d}{2}} Q^{\binom{d}{2}-i} C_i^2(q_1(P), \ldots, q_{d-1}(P)), \\[2em] \displaystyle (R-Q) \sum_{l=0}^{\binom{d+1}{3}} \sum_{n=0}^{l} C_n^3(q_1(P), \ldots, q_{d_1}(P)) \cdot \\[2em] \displaystyle \cdot \left(\frac{\sum_{i+j \le d-1} q_i(P) \binom{d-1-i}{j}(-1)^j P^j}{\sum_{i+j \le d-1} q_i(P) \binom{d-1-i}{j}(-1)^j Q^j} \right) R^{\binom{d+1}{3}-l} \end{array} \right)_{l-n}}.$$

Here $(\cdot)_n$ denotes the part of degree n $(P, Q, R$ have each degree 1 and $c_i(X)$ has degree i). We have $P = c_1(\mathcal{O}_{\mathbf{P}(T_X^*)}(1))$, $Q = c_1(\mathcal{O}_{\mathbf{P}(W_{d-1}^2(X))}(1))$, $R = c_1(\mathcal{O}_{\mathbf{P}(W_{d-1}^3(X))}(1))$.

Proof: By the exact sequences from lemma 3.2.9(2) we get

$$c(W_{d-1}^3(X)) = c(\mathrm{Sym}^3(\tilde{\pi}_2^*(Q_1)))c(\tilde{\pi}_2^*(Q_1) \otimes \tilde{\pi}_2^*(T_1))c(T_2)/c(T_2 \otimes \tilde{\pi}_2^*(Q_1)),$$

and for a vector bundle E of rank r and a line bundle A we have

$$c(E \otimes A) = \sum_{i+j=r} \binom{r-i}{j} c_i(E) c_1(A)^j. \qquad \square$$

We will rewrite these formulas explicitely for $d \leq 3$.

If X is a surface, then

$$A^*(D_1^3(X)) = \cfrac{A^*(X)[P, Q, R]}{\begin{pmatrix} P^2 + c_1(X)P + c_2(X), \\ Q^2 + (c_1(X) - P)Q + 2c_2(X), \\ R^2 + (c_1(X) - 2P)R + c_2(X) - 2c_1(X)P - PQ \end{pmatrix}}.$$

If $d = dim(X) = 3$, then

$$A^*(D_1^3(X)) = \cfrac{A^*(X)[P, Q, R]}{\begin{pmatrix} P^3 + c_1(X)P^2 + c_2(X)P + c_3(X), \\ Q^3 + (c_1(X) - P)Q^2 + (c_2(X) - c_1(X)P - P^2)Q + 2c_3(X), \\ R^3 + (c_1(X) - 2P)R^2 + (c_2(X) - 2c_1(X)P - PQ)R \\ \qquad + c_3(X) - 2c_2(X)P - c_1(X)PQ - P(Q^2 - PQ). \end{pmatrix}}$$

·and

$$A^*(D_2^3(X)) = \frac{A^*(X)[P, Q, R]}{(S_1, S_2, S_3)},$$

where

$$S_1 := P^3 - c_1(X)P^2 + c_2(X)P - c_3(X),$$
$$S_2 := (Q - P)(Q + P - c_1(X))(Q^2 + 2(P - c_1(X))Q$$
$$\qquad + 4(P^2 - c_1(X)P + c_2(X)),$$

$$S_3 := R^5 + (4P + Q - 6c_1(X))R^4 + (12P^2 + 2PQ - 23Pc_1(X)$$
$$\qquad + Q^2 - 3Qc_1(X) + 11c_1^2 + 10c_2)R^3$$

$$+ (4P^2Q - 35P^2c_1(X) - PQ^2 - 2PQc_1(X) + P(41c_1(X)^2 + 23c_2(X))$$
$$+ Q^3 + 2Q^2c_1(X) + Q(-4c_1(X)^2 + 8c_2(X)) - 6c_1(X)^3$$
$$- 30c_1(X)c_2(X) + 11c_3(X))R^2$$
$$+ (-P^2Q^2 + 9P^2Qc_1(X) + P^2(24c_1(X)^2 + 15c_2(X)) - 3PQ^3$$
$$+ 14PQ^2c_1(X) + PQ(-21c_1(X)^2 - 25c_2(X)) + P(-24c_1(X)^3$$
$$- 57c_1(X)c_2(X) + 6c_3(X)) + 6Q^3c_1(X) + Q^2(-18c_1(X)^2 + 9c_2(X))$$
$$+ Q(12c_1(X)^3 + 3c_1(X)c_2(X) - 13c_3(X)))R$$
$$- 21P^2Q^3 + 99P^2Q^2c_1(X) + P^2Q(-66c_1(X)^2 - 19c_2(X))$$
$$+ P^2(-168c_1(X)c_2(X) + 56c_3(X)) + 75PQ^3c_1(X)$$
$$+ PQ^2(-137c_1(X)^2 - 126c_2(X)) + PQ(66c_1(X)^3 + 99c_1(X)c_2(X) + 6c_3(X))$$
$$+ Q^3(-68c_1(X)^2 + 5c_2(X)) + Q^2(36c_1(X)^3 + 111c_1(X)c_2(X) - 54c_3(X)).$$

Let E be a vector bundle of rank r on a smooth projective variety X. Now we want to study the vector bundles $(E)_m^n$ from definition 3.2.6. For this purpose we first consider the bundles $\widetilde{E}_l$ on the Hilbert scheme $X^{[l]}$. We can associate in a natural way to each section s of E a section $\widetilde{s}_l$ of $\widetilde{E}_l$ and thus also a section $(s)_m^n$ of $(E)_m^n$:

Definition 3.2.12. For any point $Z \in X^{[l]}$ the fibre $\widetilde{E}_l(Z)$ of $\widetilde{E}_l$ over Z is the vector space $H^0(Z, E \otimes_{\mathcal{O}_X} \mathcal{O}_Z)$. Let

$$ev_Z : H^0(X, E) \longrightarrow H^0(Z, E \otimes_{\mathcal{O}_X} \mathcal{O}_Z)$$

be the evaluation morphism. For any section $s \in H^0(X, E)$ we define a section $\widetilde{s}_l$ of $\widetilde{E}_l$ by

$$\widetilde{s}_l(Z) := ev_Z(s)$$

and put $(s)_m^n := r_n^*(\widetilde{s}_{\binom{m+n}{n}})$. This defines the evaluation morphism

$$ev_E : H^0(X, E) \otimes \mathcal{O}_{D_m^n(X)} \longrightarrow (E)_m^n.$$

Remark 3.2.13. Let s be a section of E and $Y \subset X$ its zero locus. From definition 3.2.12 we see immediately that $Y^{[n]} \subset X^{[n]}$ is exactly the zero locus of $\widetilde{s}_l$, and thus $D_m^n(Y) \subset D_m^n(X)$ is the zero locus of $(s)_m^n$. To begin with this is only true set-theoretically, i.e. without considering the possible non-reduced structure. If $m = 1$ and $n \leq 3$ or $n \leq 2$ one can however show by computations in local coordinates on $D_m^n(X)$ that the smooth subvariety $D_m^n(Y)$ is the zero locus of $(s)_m^n$ in $D_m^n(X)$, if

Y is smooth of codimension r. In particular we see in this case for the top Chern classes

$$c_r(E) = [Y] \in A^r(X),$$

$$c_{r\binom{m+n}{n}}((E)_m^n) = [D_m^n(Y)] \in A^{r\binom{m+n}{n}}(D_m^n(X)).$$

The vector bundles $(E)_m^n$ can be related to the simplest case $E = \mathcal{O}_X$: let $\Delta \subset X \times X$ be the diagonal, $\mathcal{I}_\Delta$ its ideal and $\Delta^{n+1} \subset X \times X$ the subscheme defined by $\mathcal{I}^{n+1}$. Let $\mathrm{Hilb}^l(\Delta^{n+1}/X) \subset \mathrm{Hilb}^l(X \times X/X) = X^{[l]}$ be the relative Hilbert scheme of subschemes of length n of Δ^{n+1} over X. Let $\pi : \mathrm{Hilb}^l(\Delta^{n+1}/X) \longrightarrow X$ and $r_0 : D_m^n(X) \longrightarrow X$ be the projections.

Lemma 3.2.14. *Let E be a vector bundle on E.*

(1) $\tilde{E}_l|_{\mathrm{Hilb}^l(\Delta^{n+1}/X)} = \pi^*(E) \otimes \left((\tilde{\mathcal{O}}_X)_l|_{\mathrm{Hilb}^l(\Delta^{n+1}/X)}\right).$

(2) *For all $n \in \mathbb{N}$, $m \leq d$ we have $(E)_m^n = r_0^*(E) \otimes (\mathcal{O}_X)_m^n$.*

Proof: (1) For all $l \in \mathbb{N}$ we put $Z_l(X)_{(n)} := \mathrm{Hilb}^l(\Delta^{n+1}/X) \times_{X^{[l]}} Z_l(X)$. Let

$$Z_l(X)_{(n)}$$

$$\swarrow q \qquad\qquad \searrow p$$

$$X \qquad\qquad\qquad \mathrm{Hilb}^l(\Delta^{n+1}/X)$$

be the projections. Then we have the commutative diagram

$$
\begin{array}{ccc}
\mathrm{Hilb}^l(\Delta^{n+1}/X) & \overset{p}{\longleftarrow} & Z_l(X)_{(n)} \\
\downarrow{\scriptstyle\pi} & & \downarrow{\scriptstyle q} \\
X & == & X,
\end{array}
$$

and by the projection formula we get

$$
\begin{aligned}
\tilde{E}_l|_{\mathrm{Hilb}^l(\Delta^{n+1}/X)} &= p_*(q^*(E)) \\
&= p_*(p^*(\pi^*(E))) \\
&= \pi^*(E) \otimes (\tilde{\mathcal{O}}_X)_l|_{\mathrm{Hilb}^l(\Delta^{n+1}/X)}.
\end{aligned}
$$

So we get (1). The projection

$$r_n : D_m^n(X) \longrightarrow X^{\left[\binom{m+n}{n}\right]}$$

factors through $\mathrm{Hilb}^{\binom{n+m}{m}}(\Delta^{n+1}/X)$ (see the remarks after definition 3.2.1). So (2) follows from (1). □

Now we specialize to the case $X = \mathbf{P}_N$ and to the hyperplane bundle $H = \mathcal{O}(1)$.

Proposition 3.2.15.

(1) *Let*

$$H := c_1(\mathcal{O}_{\mathbf{P}_N}(1)),$$
$$P := c_1(\mathcal{O}_{\mathbf{P}(T_{\mathbf{P}_N})}(1)),$$
$$Q := c_1(\mathcal{O}_{\mathbf{P}(W_1^2(\mathbf{P}_N)^*)}(1)),$$
$$R := c_1(\mathcal{O}_{\mathbf{P}(W_1^3(\mathbf{P}_N)^*)}(1)).$$

Then we have in $A^*(D_1^2(\mathbf{P}_N))$

$$c((H)_1^2) = (1 + (3H + P + Q) + (3H^2 + 2H(P + Q) + PQ)$$
$$+ (H^3 + H^2(P + Q) + HPQ)$$

and in $A^*(D_1^3(\mathbf{P}_N))$

$$c((H)_1^3) = 1 + (4H + P + Q + R)$$
$$+ (6H^2 + 3H(P + Q + R) + PQ + PR + QR)$$
$$+ (4H^3 + 3H^2(P + Q + R) + 2H(PQ + PR + QR) + PQR)$$
$$+ (H^4 + H^3(P + Q + R) + H^2(PQ + PR + QR) + HPQR).$$

(2) *Let* $d_1, \ldots, d_m$ *be the Chern classes of the universal quotient bundle on* $D_m^1(X) = Grass(m, T_{\mathbf{P}_N}^*)$ *and* $f_1, \ldots, f_{\binom{m+1}{2}}$ *the Chern classes of the universal quotient bundle on* $D_m^2(X) = Grass(\binom{m+1}{2}, W_m^2(X))$. *Then we have*

$$c((H)_m^2) = \sum_{i+j \leq \binom{m+2}{2}} \binom{\binom{m+2}{2} - i}{j} \left(\sum_{k+l=i} d_k f_l \right) H^j$$

(3) *If* $m = N-1$, *let in addition* $h_1, \ldots, h_{\binom{N+1}{3}}$ *be the Chern classes of the universal quotient bundle on*

$$D_{N-1}^3(X) = Grass(\binom{N+1}{3}, W_{N-1}^3(X)).$$

Then we have

$$c((H)_{N-1}^3) = \sum_{i+j \leq \binom{N+2}{3}} \binom{\binom{N+2}{3} - i}{j} \left(\sum_{k+l+s=i} d_k f_l h_s \right) H^j.$$

Here in (2) *and* (3) *we formally set* $d_k = 0$ *for* $k > m$, $f_l = 0$ *for* $l > \binom{m+1}{2}$ *and* $h_s = 0$ *for* $s > \binom{N+1}{3}$.

Proof: This follows from lemma 3.2.14 and the exact sequences

$$0 \longrightarrow Q_1 \longrightarrow (\mathcal{O}_X)^1_m \longrightarrow \mathcal{O}_X \longrightarrow 0$$
$$0 \longrightarrow Q_2 \longrightarrow (\mathcal{O}_X)^2_m \longrightarrow (\mathcal{O}_X)^1_m \longrightarrow 0$$
$$0 \longrightarrow Q_3 \longrightarrow (\mathcal{O}_X)^3_m \longrightarrow (\mathcal{O}_X)^2_m \longrightarrow 0 \quad \square$$

Now we want to compute the class $[D_1^2(C)] \in A^{3N-3}(D_1^2(\mathbf{P}_N))$ for a smooth curve $C \subset \mathbf{P}_N$.

Proposition 3.2.16.

$$[D_1^2(C)] = deg(C)H^{N-1}P^{N-1}Q^{N-1}$$
$$+ ((N+1)deg(C) + 2g(C) - 2)(H^N P^{N-2}Q^{N-1} + H^N P^{N-1}Q^{N-2})$$

Proof: We have $H \cdot [D_1^2(C)] = deg(C)$. By remark 3.2.8 we also have

$$P \cdot [D_1^2(C)] = 2g(C) - 2,$$
$$Q \cdot [D_1^2(C)] = 4g(C) - 4.$$

On the other hand we can use the relations to compute the intersection table:

	$H^N P^{N-1}Q^{N-2}$	$H^N P^{N-2}Q^{N-1}$	$H^{N-1}P^{N-1}Q^{N-1}$
H			1
P		1	$-N-1$
Q	1	1	$-2N-2$

This proves the result. $\square$

Enumerative applications for contacts of projective varieties with linear subvarieties of $\mathbf{P}_N$

Now we want to apply our considerations to obtain formulas for the numbers of higher order contacts of a smooth projective variety $X \subset \mathbf{P}_N$ of dimension d with linear subvarieties of $\mathbf{P}_N$ of dimension m. We have to distiguish two cases: $m \geq d$ and $m < d$. We will see that the first case is the simpler one, as in this case we have $X = D_m^n(X)$, and so the computations can be carried out directly in the Chow ring of X. In case $m < d$ we have to consider the more complicated Chow rings of $D_m^2(X)$ and $D_1^3(X)$.

We again want to use the Porteous formula. Let $H = \mathcal{O}_{\mathbf{P}_N}(1)$ be the hyperplane bundle on $\mathbf{P}_N$. We will denote by the same letter its restriction to X and its first Chern class.

Contacts with linear subvarieties of higher dimension

Let $X \subset \mathbf{P}_N$ be a smooth m-dimensional subvariety. We can in a natural way identify $D_m^n(X)$ with X for all $n \in \mathbb{N}$, and with this identification we get

$$(H)_m^n = H \otimes (\mathcal{O}_X)_m^n = H \otimes J_n(X).$$

On $X = D_m^n(X)$ we consider the evaluation morphism

$$ev_m : H^0(\mathbf{P}_N, \mathcal{O}_{\mathbf{P}_N}(1)) \otimes \mathcal{O}_X \longrightarrow (H)_m^n.$$

This is the composition of the restriction

$$r : H^0(\mathbf{P}_N, \mathcal{O}_{\mathbf{P}_N}(1)) \longrightarrow H^0(X, H)$$

with the evaluation morphism

$$ev_H : H^0(X, H) \otimes \mathcal{O}_X \longrightarrow (H)_m^n$$

from definition 3.2.12. Over every point $x \in X$ the kernel of the induced map

$$ev_m(x) : H^0(\mathbf{P}_N, \mathcal{O}_{\mathbf{P}_N}(1)) \longrightarrow H^0\big(spec(\mathcal{O}_{X,x}/\mathrm{m}_{X,x}^{n+1}), H \otimes (\mathcal{O}_{X,x}/\mathrm{m}_{X,x}^{n+1})\big)$$

on the fibres consists of the sections $s \in H^0(\mathbf{P}_N, \mathcal{O}_{\mathbf{P}_N}(1))$ for which the hyperplane $\mathbf{P}(ker(s)) \subset \mathbf{P}(H^0(\mathbf{P}_N, \mathcal{O}_{\mathbf{P}_N}(1))^*)$ has n^{th} order contact with X. A linear subvariety V of $\mathbf{P}_N$ of dimension $m_1 \geq m$ has n^{th} order contact in x, if and only if each hyperplane of $\mathbf{P}_N$ containing V has n^{th} order contact with X at x. So the locus where X has n^{th} order contact with an l-codimensional linear subvariety of $\mathbf{P}_N$ is the degeneracy locus

$$\mathcal{D}_{N+1-l}(ev_m) = \big\{ x \in X \mid rk(ev_m(x)) \leq N + 1 - l \big\}.$$

So we get by Porteous formula (see. theorem 3.1.18):

Proposition 3.2.17. *Let* $X \subset \mathbf{P}_N$ *be a smooth closed subvariety of dimension* m. *The locus where* X *has* n^{th} *order contact with* l-*codimensional linear subvarieties of* $\mathbf{P}_N$ *has at most codimension*

$$r := l \cdot \left(\binom{m+n}{n} - N - 1 + l \right)$$

in X. *If its codimension is* r, *then its class is*

$$\det\left(\left(c_{\binom{m+n}{n}-N-1+l+i-j}(J_n(X) \otimes H) \right)_{1 \leq i,j \leq l} \right) \in A^r(X).$$

In particular we have:

(1) *The class of the locus, where* X *has* n^{th} *order contact with a hyperplane in* $\mathbf{P}_N$ *is*

$$\sum_{i+j=\binom{m+n}{n}-N} \sum_{i_1+\ldots+i_n=i} (-1)^i \binom{\binom{n+m}{n}-i}{j} \prod_{l=1}^{n} c_{i_l}(\text{Sym}^l(T_X)) H^j.$$

(2) *Let* $C \subset \mathbf{P}_n$ *be a smooth curve. The number of* n^{th} *order contacts of* C *with hyperplanes in* $\mathbf{P}_n$ *is*

$$\binom{n+1}{2}(2g(C)-2) + (n+1)\deg(C).$$

(3) *Let* S *be a smooth surface in* $\mathbf{P}_N$. *If* $N = \binom{n+1}{2} - 1$, *then the class of the* $(n-1)^{th}$ *order contacts with hyperplanes in* $\mathbf{P}_N$ *is*

$$-\sum_{k=1}^{n-1} k(n-k) c_1(S) + \binom{n+1}{2} H.$$

If $N = \binom{n+1}{2} - 2$, *then the number of* $(n-1)^{th}$ *order contacts with hyperplanes in* $\mathbf{P}_N$ *is*

$$\sum_{m=1}^{n-1} \sum_{0 \leq k < \frac{m}{2}} (m-2k)^2 c_2(S)$$

$$+ \left(\sum_{i=2}^{n-1} \binom{i}{2}(n-i)^2 + \sum_{1 \leq i < j \leq n-1} ij(n-i)(n-j) \right) c_1(S)^2$$

$$- \sum_{k=1}^{n-1} k(n-k) \left(\binom{n+1}{2} - 1 \right) c_1(S)H + \binom{\binom{n+1}{2}}{2} H^2.$$

If $N = \binom{n+1}{2}$, then the number of $(n-1)^{th}$ order contacts with 2-codimensional linear subspaces in $\mathbf{P}_N$ is

$$\left(\sum_{1 \le i \le j \le n-1} ij(n-i)(n-j) - \sum_{i=2}^{n-1} \binom{i}{2}(n-i)^2 \right) c_1(S)^2$$

$$- \sum_{m=1}^{n-1} \sum_{0 \le k < \frac{m}{2}} (m-2k)^2 c_2(S)$$

$$- \sum_{k=1}^{n-1} k(n-k)\left(\binom{n+1}{2} + 1 \right) c_1(S)H + \left(\binom{\binom{n+1}{2}+1}{2} \right) H^2.$$

(4) *Let X be a smooth threefold in $\mathbf{P}_9$. The class of second order contacts of X with hyperplanes is*

$$-5c_1(X) + 10H.$$

Let X be a smooth threefold in $\mathbf{P}_8$. The class of second order contacts of X with hyperplanes is

$$9c_1(X)^2 + 6c_2(X) - 45c_1(X)H + 45H^2.$$

Let X be a smooth threefold in $\mathbf{P}_7$. The number of second order contacts of X with hyperplanes is

$$-7c_1(X)^3 - 20c_1(X)c_2(X) - 8c_3(X) + 72c_1(X)^2 H$$
$$+ 48c_2(X)H - 180c_1(X)H^2 + 120H^3.$$

Let X be a smooth threefold in $\mathbf{P}_{10}$. The class of second order contacts of X with 2-codimensional linear subvarieties is

$$16c_1(X)^2 - 6c_2(X) - 55c_1(X)H + 55H^2.$$

The number of second order contacts of X with 3-codimensional linear subvarieties is

$$-42c_1(X)^3 + 40c_1(X)c_2(X) - 8c_3(X) + 192c_1(X)^2 H$$
$$- 72c_2(X)H - 330c_1(X)H^2 + 220H^3.$$

Obviously (1)-(4) only hold in the case that the locus where the contact occurs has the right codimension in X.

Proof: By $(H)_m^n = J_n(X) \otimes H$ the total Chern class satisfies

$$c((H)_m^n) = \sum_{i+j \le \binom{m+n}{n}} \sum_{i_1 + \ldots + i_n = i} (-1)^i \binom{\binom{n+m}{n} - i}{j} \prod_{l=1}^{n} c_{i_l}(\mathrm{Sym}^l(T_X))H^j.$$

From this we immediately get (1). (2) follows by an easy computation.

(3) By (1) and remark 3.1.9 the coefficients of $c_1(X)$ and $c_1(X)^2$ in $c((H)_2^{n-1})$ are the coefficients of x_1 and x_1^2 in

$$\prod_{k=1}^{n-1}(1 - (n - k)x_1)^k$$

respectively, and the coefficient of $c_2(X)$ is the number

$$\sum_{m=2}^{n-1} \sum_{0 \leq k < \frac{m}{2}} (m - 2k)^2.$$

The rest follows by an easy computation. (4) follows from (1) and remark 3.1.9 by an easy computation. $\square$

Contacts with linear subvarieties of lower dimension

Let $X \subset \mathbf{P}_N$ be a smooth projective variety of dimension d. Now we want to treat the second order contacts of X with linear subvarieties of $\mathbf{P}_N$ of dimensions $m < d$ and also the third order contacts of X with lines. We first study the case of second order contacts. On $D_m^2(X)$ we consider the evaluation morphism

$$ev_m : H^0(\mathbf{P}_N, \mathcal{O}_{\mathbf{P}_N}(1) \otimes \mathcal{O}_{D_m^2(X)} \longrightarrow (H)_m^2.$$

This is the composition of the restriction

$$r : H^0(\mathbf{P}_N, \mathcal{O}_{\mathbf{P}_N}(1)) \longrightarrow H^0(X, H)$$

with the evaluation morphism

$$ev_H : H^0(X, H) \otimes \mathcal{O}_{D_m^2(X)} \longrightarrow (H)_m^2.$$

Over each point $w = (x, Z_1, Z_2) \in D_m^2(X)$ the kernel of the induced map

$$ev_m(w) : H^0(\mathbf{P}_N, \mathcal{O}_{\mathbf{P}_N}(1)) \longrightarrow H^0(Z_2, H \otimes \mathcal{O}_{Z_2})$$

on the fibres consists of the sections $s \in H^0(\mathbf{P}_N, \mathcal{O}_{\mathbf{P}_N}(1))$ for which the hyperplane $\mathbf{P}(ker(s)) \subset \mathbf{P}(H^0(\mathbf{P}_N, \mathcal{O}_{\mathbf{P}_N}(1))^*)$ contains Z_2 as a subscheme. A linear subvariety V of $\mathbf{P}_N$ of dimension m contains Z_2 as a subscheme, if and only if each hyperplane containing V also contains Z_2. So the locus

$$\left\{ w = (x, Z_1, Z_2) \in D_m^2(X) \,\middle|\, Z_2 \text{ lies on an } m\text{-plane} \right\}$$

is exactly the degeneracy locus

$$\mathcal{D}_{m+1}(ev_m) = \Big\{ w \in D_m^2(X) \ \Big| \ rk(ev_m(w)) \le m+1 \Big\}.$$

Let $r_0 : D_m^2(X) \longrightarrow X$ be the projection. From the above we get for the image of the degeneracy locus

$$r_0(\mathcal{D}_{m+1}(ev_m)) = \left\{ x \in X \ \middle| \ \begin{array}{c} \text{there is an } m\text{-plane} \\ \text{having second order contact with } X \text{ in } x \end{array} \right\}.$$

So $(r_0)_*(\mathcal{D}_{m+1}(ev_m)) \in A^*(X)$ is the class of the locus where X has second order contact with m-planes counted with multiplicities. Let W be an irreducible component of $r_0(\mathcal{D}_{m+1}(ev_m))$. The multiplicity of W in $(r_0)_*(\mathcal{D}_{m+1}(ev_m))$ is the degree of $r_0|_{D_{m+1}(ev_m)}$ over W (or zero if this degree is infinite), i.e. the number of m-planes having second order contact in a general point of W counted with multiplicities. So we call $(r_0)_*(\mathcal{D}_{m+1}(ev_m))$ the class of second order contacts of X with m-planes in $\mathbf{P}_N$.

We can also determine this class in a dual way:

let

$$ev_m^* : ((H)_m^2)^* \longrightarrow (H^0(\mathbf{P}_N, \mathcal{O}_{\mathbf{P}_N}(1)) \otimes \mathcal{O}_{D_m^2(X)})^*$$

be the dual morphism of ev_m. For $w = (x, Z_1, Z_2) \in D_m^2(X)$ the subscheme Z_2 lies on an m-plane if and only if $ev_m^*(w)$ has at most rank $m+1$. So the set

$$\Big\{ w = (x, Z_1, Z_2) \in D_m^2(X) \ \Big| \ Z_2 \text{ lies on an } m\text{-plane} \Big\}$$

is the degeneracy locus $\mathcal{D}_{m+1}(ev_m^*)$. So we get:

Proposition 3.2.18. *Let X be a smooth projective variety of dimension d in $\mathbf{P}_N$. If the locus where X has second order contact with m-planes has codimension at least*

$$r := (N - d)\binom{m+1}{2} - (d-m)m,$$

then its class is

$$(r_0)_* \Big(det \Big(\Big(c_{\binom{m+1}{2}+i-j}((H)_m^2) \Big)_{1 \le i,j \le N-m} \Big) \Big)$$

$$= (r_0)_* \Big(det \Big(\Big(s_{N-m+i-j}(((H)_m^2)^*) \Big)_{1 \le i,j \le \binom{m+1}{2}} \Big) \Big) \in A^r(X).$$

In particular the class of second order contacts of X with lines is

$$(r_0)_*(s_{N-1}(((H)_1^2)^*)) \in A^{N-2d+1}(X),$$

if this locus has codimension $N - 2d + 1$.

In a similar way we can argue for third order contacts with lines. Let $X \subset \mathbf{P}_N$ be a smooth projective variety. On $D_1^3(X)$ we consider the evaluation morphism

$$ev : H^0(\mathbf{P}_N, \mathcal{O}_{\mathbf{P}_N}(1)) \otimes \mathcal{O}_{D_1^3(X)} \longrightarrow (H)_1^3.$$

Let

$$ev^* : ((H)_1^3)^* \longrightarrow (H^0(\mathbf{P}_N, \mathcal{O}_{\mathbf{P}_N}(1)))^* \otimes \mathcal{O}_{D_1^3(X)}$$

be the dual morphism. For $w = (x, Z_1, Z_2, Z_3) \in D_m^2(X)$ the subscheme Z_3 lies on a line $l \subset \mathbf{P}_N$, if and only if $ev^*(w)$ has rank 2. So the locus of third order contacts of X with lines in $\mathbf{P}_N$ is the degeneracy locus $\mathcal{D}_2(ev^*)$. Let $r_0 : D_1^3(X) \longrightarrow X$ be the projection. Then we get as above:

Proposition 3.2.19. *Let* $X \subset \mathbf{P}_N$ *be a smooth variety of dimension* d. *If the codimension of the locus, where* X *has third order contact with lines, has codimension* $2N - 3d + 1$, *then its class is*

$$(r_0)_*(s_{N-1}(((H)_m^3)^*)^2 - s_N(((H)_m^3)^*)s_{N-2}(((H)_m^3)^*)) \in A^{2N-3d+1}(X).$$

As we know the Chow rings of $D_m^2(X)$ and $D_1^3(X)$, and the Chern classes of $(H)_m^2$ and $(H)_1^3$ can be expressed in terms of the generators of these cohomology rings, we can in principle compute the classes of second order contacts with m-planes and the classes of third order contacts with lines. Note however that the Chow ring of $D_m^2(X)$ is quite complicated for $m \geq 2$. For the explicit computation we will therefore restrict ourselves to the case of contacts with lines. We compute these classes for small N with the help of a computer. The total Segre class of $((H)_1^2)^*$ is

$$s(((H)_1^2)^*) := (1 - H)^{-1}(1 - (P + H))^{-1}(1 - (Q + H))^{-1},$$

and the total Segre class of $((H)_1^3)^*$ is

$$s(((H)_1^3)^*) := (1 - H)^{-1}(1 - (P + H))^{-1}(1 - (Q + H))^{-1}(1 - (R + H))^{-1}.$$

So we get the following formula:

The class of second order contacts of a smooth surface $X \subset \mathbf{P}_4$ with lines is

$$2(-3c_1(X) + 5H).$$

The number of second order contacts of a smooth surface $X \subset \mathbf{P}_5$ with lines is

$$2(7c_1(X)^2 - 5c_2(X) - 18c_1(X)H + 15deg(X)).$$

This formula has been obtained in [Le Barz (4),(9)] using a different method.

The class of second order contacts of a smooth threefold $X \subset \mathbf{P}_6$ with lines is

$$4(-3c_1(X) + 7H).$$

The class of second order contacts of a smooth threefold $X \subset \mathbf{P}_7$ with lines is

$$4(7c_1(X)^2 - 5c_2(X) - 24c_1(X)H + 28H^2).$$

The number of second order contacts of a smooth threefold $X \subset \mathbf{P}_8$ with lines is

$$12(-5c_1(X)^3 + 8c_1(X)c_2(X) - 3c_3(X) + 21c_1(X)^2H - 15c_2(X)H$$
$$- 36c_1(X)H^2 - 28deg(X)).$$

The class of second order contacts of a smooth fourfold $X \subset \mathbf{P}_9$ with lines is

$$8(7c_1(X)^2 - 5c_2(X) - 30c_1(X)H + 45H^2).$$

The class of second order contacts of a smooth fourfold $X \subset \mathbf{P}_{10}$ with lines is

$$8(-15c_1(X)^3 + 24c_1(X)c_2(X) - 9c_3(X) + 77c_1(X)^2H - 55c_2(X)H$$
$$- 165c_1(X)H^2 + 165H^3).$$

The number of second order contacts of a smooth fourfold $X \subset \mathbf{P}_{11}$ with lines is

$$8(31c_1(X)^4 - 79c_1(X)^2c_2(X) + 21c_2(X)^2 + 44c_1(X)c_3(X) - 17c_4(X)$$
$$- 180c_1(X)^3H + 288c_1(X)c_2(X)H - 108c_3(X)H$$
$$+ 462c_1(X)^2H^2 - 330c_2(X)H^2 - 660c_1(X)H^3 + 495deg(X)).$$

The class of third order contacts of a smooth threefold $X \subset \mathbf{P}_5$ with lines is

$$85c_1(X)^2 - 49c_2(X) - 330c_1(X)H + 411H^2.$$

The class of third order contacts of a smooth fourfold $X \subset \mathbf{P}_7$ with lines is

$$-575c_1(X)^3 + 790c_1(X)c_2(X) - 251c_3(X) + 3400c_1(X)^2H$$
$$- 1960c_2(X)H - 8228c_1(X)H^2 + 8680H^3.$$

In section 3.3 we will develop a new method of determining a formula for higher order contacts of a smooth variety $X \subset \mathbf{P}_N$ with lines in $\mathbf{P}_N$. At the end we will obtain a general formula which contains the ones above as special cases.

We briefly want to consider the contacts of a projective variety with more general families of subvarieties of $\mathbf{P}_N$.

Definition 3.2.20. Let T be a smooth projective variety and $Y \longrightarrow T$ a smooth morphism of relative dimension m. Here we asume Y to be quasiprojective over T. We put $\tilde{D}^1_m(Y/T) := Grass(m, \Omega_{Y/T})$. We define the vector bundle $W^2_m(Y/T)$ on $\tilde{D}^1_m(Y/T)$ in an analogous way to $W^2_m(X)$, replacing the bundles by their relative versions relative to T. Then we put $\tilde{D}^2_m(Y/T) := Grass\left(\binom{m+1}{2}, W^2_m(Y/T)\right)$.

It is obvious from the definitions, that both $\tilde{D}^1_m(Y/T)$ and $\tilde{D}^2_m(Y/T)$ are isomorphic to Y.

Definition 3.2.21. Let T be a smooth variety and $Y_T \subset \mathbf{P}_N \times T$ a flat family of m-dimensional subvarieties of $\mathbf{P}_N$, i.e. we have the projections

$$Y_T$$
$$\swarrow p_1 \qquad \searrow p_2$$
$$\mathbf{P}_N \qquad\qquad\qquad T$$

p_2 is flat, and for all $t \in T$ the fibre $Y_t = p_2^{-1}(t)$ has pure dimension m. In addition we assume that Y_T is irreducible, and there is a dense open subset $Y_{T,0} \subset Y_T$ such that the restriction $Y_{T,0} \longrightarrow T$ is a smooth morphism.

Then $\tilde{D}^2_m(Y_{T,0}/T)$ is a locally closed subvariety of

$$\tilde{D}^2_m((\mathbf{P}_N \times T)/T) = \tilde{D}^2_m(\mathbf{P}_N) \times T = D^2_m(\mathbf{P}_N) \times T,$$

if we again identify $D^2_m(\mathbf{P}_N)$ and $\tilde{D}^2_m(\mathbf{P}_N)$ via ϕ_2. Let $\bar{D}^2_m(Y_T)$ be the closure of $\tilde{D}^2_m(Y_{T,0}/T)$ in $D^2_m(\mathbf{P}_N) \times T$ and $[\bar{D}^2_m(Y_T)]$ its class in $A^*(D^2_m(\mathbf{P}_N) \times T)$. Let

$$p : D^2_m(\mathbf{P}_N) \times T \longrightarrow D^2_m(\mathbf{P}_N)$$

be the projection. Let $X \subset \mathbf{P}_N$ be a smooth projective variety of dimension $d \geq m$. Let $i : D^2_m(X) \longrightarrow D^2_m(\mathbf{P}_N)$ be the embedding and $r_0 : D^2_m(X) \longrightarrow X$ be the projection. We put

$$K(X, Y_T) := (r_0)_*(i^*(p_*([\bar{D}^2_m(Y_T)]) \in A^*(X).$$

Remark 3.2.22. $K(X, Y_T)$ is a candidate for the class of the locus where X and elements of the family Y_T have second order contact.

Proposition 3.2.23. *Let $n, d \in I\!N$. Let $Y_T \subset \mathbf{P}_N \times T$ be a family of m-dimensional projective varieties satisfying the conditions of definition 3.2.21 with $\dim(T) = t$.*

Let $e = (N - d)\binom{m+2}{2} - t + (d - m)$, *and assume* $0 \le e \le d$. *For all partitions* $(\alpha) = (1^{\alpha_1}, 2^{\alpha_2}, \ldots)$ *of numbers* $s \le e$ *there are integers* n_α *such that we have for all smooth projective varieties* $X \subset \mathbf{P}_N$ *of dimension* d:

$$K(X, Y_T) = \sum_{s=0}^{e} \sum_{\alpha \in P(s)} n_\alpha H^{e-s} c_1(X)^{\alpha_1} \ldots c_e(X)^{\alpha_e}.$$

Proof: Let $f := \binom{m+2}{2}(N - m) - t$. We will show more generally that for every class $W \in A^f(D_m^2(\mathbf{P}_N))$ there are integers n_α for all partitions α of numbers $s \le e$ such that the above formula holds for $(r_0)_*(i^*(W))$. As $A^*(D_m^2(\mathbf{P}_N))$ is generated by H and the Chern classes of the universal quotient bundles Q_1 and Q_2, it is enough to show the result for the monomials M in H and the Chern classes of Q_1 and Q_2. Using our conventions we can write $i^*(H) = H$, $i^*(Q_1) = Q_1$ and $i^*(Q_2) = Q_2$. Let $M = M_0 M_1 M_2$, where M_0 is a monomial in H and the Chern classes of X, M_1 a monomial in the Chern classes of Q_1 and M_2 a monomial in the Chern classes of Q_2. We assume that $M_1 \in A^{d_1}(D_m^2(X))$, $M_2 \in A^{d_2}(D_m^2(X))$. If $d_1 = m(d - m)$ and $d_2 = \binom{m+1}{2}(d - m)$, then we have $(r_0)_*(M) = a M_0$ for a suitable integer a depending only on the monomials M_1 and M_2 and not on X. (Let $q_1, \ldots, q_m$ and $r_1, \ldots, r_{\binom{m+1}{2}}$ be the Chern classes of the universal quotient bundle on $Grass(m, d)$ and on $Grass(\binom{m+1}{2}, \binom{m+1}{2} + d - m)$ respectively. Then a is the product of the intersection numbers $M_1(q_1, \ldots, q_m)$ and $M_2(r_1, \ldots, r_{\binom{m+1}{2}})$ on these Grassmannians.) If $d_2 < \binom{m+1}{2}(d - m)$ or $d_2 = \binom{m+1}{2}(d - m)$ and $d_1 < m(d - m)$, then we have $(r_0)_*(M) = 0$. If $d_2 > \binom{m+1}{2}(d - m)$, then we use the relations of propositions 3.1.11 to express M as a linear combination with $\mathbb{Z}$-coefficients of monomials $N = N_0 N_1 N_2$, where $N_2 \in A^{e_2}(D_m^2(X))$ with $e_2 < d_2$. If $d_2 = \binom{m+1}{2}(d - m)$ and $d_1 > m(d - m)$, then we use the relations of proposition 3.1.11 to express M as a linear combination with $\mathbb{Z}$-coefficients of monomials $N = N_0 N_1 M_2$, where $N_1 \in A^{e_1}(D_m^2(X))$ with $e_1 < d_1$. So the result follows by induction. $\square$

3.3. Semple bundles and the formula for contacts with lines

In this section we introduce the Semple bundle varieties $F_n(X)$ of a smooth variety X. They parametrize in a slightly different sense than $D_1^n(X)$ the nth order data of curves on X. Like $D_1^2(X)$ and $D_1^3(X)$ they are smooth compactifications of $X_{(n+1),c}^{[n+1]}$ by a tower of $\mathbf{P}_{d-1}$-bundles over X ($d = dim(X)$). Remember that $X_{(n+1),c}^{[n+1]}$ parametrizes the nth order data of germs of smooth curves on X. We will use the $F_n(X)$ to obtain a general formula for the higher order contacts of a smooth variety $X \subset \mathbf{P}_N$ with lines in $\mathbf{P}_N$ as a linear combination of monomials in the hyperplane section H and the Chern classes of X. We finish by considering more generally higher order contacts of X with a family of curves.

For simplicity we will assume during the whole of section 3.3 that the ground field is $\mathbf{C}$.

Definition 3.3.1. Let X be a smooth variety of dimension d. We define inductively varieties $F_n(X)$ and vector bundles $G_n(X)$ on $F_n(X)$. Let

$$F_0(X) := X, \ G_0(X) := T_X.$$

Assume inductively that $F_0(X), \ldots, F_{n-1}(X)$ and $G_0(X), \ldots, G_{n-1}(X)$ are already defined. Assume furthermore that $G_{n-1}(X)$ is a subbundle of the tangent bundle $T_{F_{n-1}(X)}$ of rank d. Then we put

$$F_n(X) := \mathbf{P}(G_{n-1}(X)).$$

Let

$$f_{n,X} : \mathbf{P}(G_{n-1}(X)) \longrightarrow F_{n-1}(X)$$

be the projection. Let $s_n := \mathcal{O}_{\mathbf{P}(G_{n-1}(X))}(-1)$ be the tautological subbundle of $f_{n,X}^*(G_{n-1}(X))$. Let $T_{F_n(X)/F_{n-1}(X)} = (\Omega_{F_n(X)/F_{n-1}(X)})^*$ be the relative tangent bundle. We define the subbundle $G_n(X)$ of $T_{F_n(X)}$ by the diagram

$$
\begin{array}{ccccccccc}
0 & \longrightarrow & T_{F_n(X)/F_{n-1}(X)} & \longrightarrow & T_{F_n(X)} & \overset{df_{n,X}}{\longrightarrow} & f_{n,X}^*(T_{F_{n-1}(X)}) & \longrightarrow & 0 \\
 & & \| & & \uparrow & \square & \uparrow j & & \\
0 & \longrightarrow & T_{F_n(X)/F_{n-1}(X)} & \longrightarrow & G_n(X) & \longrightarrow & s_n & \longrightarrow & 0.
\end{array}
$$

Here $j : s_n \hookrightarrow f_{n,X}^*(T_{F_{n-1}(X)})$ is the composition of the natural inclusions $s_n \hookrightarrow f_{n,X}^*(G_{n-1}(X))$ and $f_{n,X}^*(G_{n-1}(X)) \hookrightarrow f_{n,X}^*(T_{F_{n-1}(X)})$. We call $G_n(X)$ the nth Semple bundle and $F_n(X)$ the nth Semple bundle variety of X.

Let the divisor $D_{n+1} \subset F_{n+1}(X)$ be defined by

$$D_{n+1} = \mathbf{P}(T_{F_n(X)/F_{n-1}(X)}) \subset \mathbf{P}(G_n(X)) = F_{n+1}(X).$$

For $0 \leq i \leq n - 1$ let

$$g_{n,X}^i := f_{i+1,X} \circ \ldots \circ f_{n,X} : F_n(X) \longrightarrow F_i(X)$$

and $g_{n,X} := g_{n,X}^0$. If this does not lead to confusion, we will not write the index X of the maps f_n, g_n^i. We put

$$F_n(X)_0 := F_n(X) \setminus \left(\bigcup_{i=2}^{n} (g_n^i)^{-1}(D_i) \right).$$

The Semple bundle varieties were first introduced in [Semple (1)]. In [Collino (1)], [Colley-Kennedy (1),(2)] they are considered for arbitrary smooth surfaces. The construction of $F_n(X)$ for an arbitrary smooth projective variety X is an obvious generalisation. For our purposes it appears to be slightly more practical to use the tangent bundles instead of the cotangent bundles in the construction.

We can easily determine an inductive formula for the Chow rings of the $F_n(X)$.

Proposition 3.3.2.

$$A^*(F_n(X)) = \frac{A^*(F_{n-1}(X)[P_n])}{\left(\sum_{i=0}^{d} c_i(G_{n-1}(X)) P_n^{d-i} \right)}.$$

Here we have $P_n = c_1(s_n^*) = c_1(\mathcal{O}_{\mathbf{P}(G_{n-1}(X))}(1))$, and the Chern classes $c_i(G_n(X))$ are computed inductively by the formula

$$c(G_n(X)) = (1 - P_n) \sum_{i+j \leq d-1} \binom{d-i}{j} f_n^*(c_i(G_{n-1}(X))) P_n^j$$

$$c(G_0(X)) = c(X).$$

Proof: This follows immediately from the exact sequence

$$0 \longrightarrow T_{F_n(X)/F_{n-1}(X)} \longrightarrow G_n(X) \longrightarrow s_n \longrightarrow 0$$

and the Euler sequence

$$0 \longrightarrow \mathcal{O}_{F_n(X)} \longrightarrow f_n^*(G_{n-1}(X)) \otimes s_n^* \longrightarrow T_{F_n(X)/F_{n-1}(X)} \longrightarrow 0.$$

$\square$

Let $Y \subset X$ be a smooth closed subvariety of codimension r. Let $N_{Y/X}$ be the normal bundle of Y in X. We now want to show that $F_n(Y)$ is a closed subvariety of $F_n(X)$, and want to describe its class in the Chow ring $A^*(F_n(X))$. We suppress $g_{n,X}^*$ and $(g_{n,X}^i)^*$ in the notation.

Lemma 3.3.3. $F_n(Y)$ *is a closed subvariety of*

$$f_{n,X}^{-1}(F_{n-1}(Y)) \subset g_{n,X}^{-1}(Y) \subset F_n(X),$$

and its class $[F_n(Y)] \in A^*(g_{n,X}^{-1}(Y))$ *is*

$$[F_n(Y)] = c_r(N_{Y/X} \otimes s_1^*) c_r(N_{Y/X} \otimes s_1^* \otimes s_2^*) \ldots c_r(N_{Y/X} \otimes s_1^* \otimes \ldots \otimes s_n^*).$$

Proof: We assume by induction that $F_n(Y)$ is a closed subvariety of $F_n(X)$. On $F_n(Y)$ we have the diagram

$$
\begin{array}{ccccccccc}
0 & \longrightarrow & T_{F_n(X)/F_{n-1}(X)}|_{F_n(Y)} & \longrightarrow & G_n(X)|_{F_n(Y)} & \longrightarrow & s_n & \longrightarrow & 0 \\
 & & \uparrow & & \uparrow & & \| & & \quad (*) \\
0 & \longrightarrow & T_{F_n(Y)/F_{n-1}(Y)} & \longrightarrow & G_n(Y) & \longrightarrow & s_n & \longrightarrow & 0
\end{array}
$$

So $F_{n+1}(Y) = \mathbf{P}(G_n(Y))$ is a closed subvariety of $f_{n+1,X}^{-1}(F_n(Y)) = \mathbf{P}(G_n(X)|_{F_n(Y)})$. To determine the class $[F_{n+1}(Y)] \in A^*(g_{n+1,X}^{-1}(Y))$ we have by induction only to determine the class of $F_{n+1}(Y)$ in $A^*(f_{n+1,X}^{-1}(F_n(Y)))$. For this we consider the canonical injection

$$\sigma : s_{n+1} \hookrightarrow f_{n+1,X}^*(G_n(X)|_{F_n(Y)})$$

on $f_{n+1,X}^{-1}(F_n(Y))$. The subvariety $F_{n+1}(Y) \subset f_{n+1,X}^{-1}(F_n(Y))$ is the locus where σ factors through the subbundle $f_{n+1,X}^*(G_n(Y))$ of $f_{n+1,X}^*(G_n(X)|_{F_n(Y)})$, i.e. the vanishing locus of the composition

$$s_{n+1} \overset{\sigma}{\longrightarrow} f_{n+1,X}^*(G_n(X)|_{F_n(Y)}) \longrightarrow f_{n+1,X}^*(G_n(X)|_{F_n(Y)}/G_n(Y)).$$

As $F_{n+1}(Y)$ has codimension r in $f_{n+1,X}^{-1}(F_n(Y))$, its class in $A^r(f_{n+1,X}^{-1}(F_n(Y)))$ is the Chern class

$$c_r(s_{n+1}^* \otimes f_{n+1,X}^*(G_n(X)|_{F_n(Y)}/G_n(Y))),$$

and by the diagram $(*)$ we have:

$$G_n(X)|_{F_n(Y)}/G_n(Y) \cong (T_{F_n(X)/F_{n-1}(X)}|_{F_n(Y)})/T_{F_n(Y)/F_{n-1}(Y)}.$$

It is well known that the relative tangent bundle of a projectivized vector bundle E of rank r is

$$T_{\mathbf{P}(E)/Y} = \mathcal{O}_{\mathbf{P}(E)}(1) \otimes Q_{r-1,E}.$$

So we have

$$G_n(X)|_{F_n(Y)}/G_n(Y)$$
$$\cong s_n^* \otimes \left((f_{n,Y}^*(G_{n-1}(X)|_{F_{n-1}(Y)}))/s_n\right)/(f_{n,Y}^*(G_{n-1}(Y))/s_n))$$
$$\cong s_n^* \otimes f_{n,Y}^*(G_{n-1}(X)|_{F_{n-1}(Y)}/G_{n-1}(Y)).$$

So we get by induction

$$G_n(X)|_{F_n(Y)}/G_n(Y) \cong s_n^* \otimes \ldots \otimes (g_{n,X}^1)^*(s_1^*) \otimes g_{n,X}^*(T_X|_Y/T_Y)$$
$$\cong s_n^* \otimes \ldots \otimes (g_{n,X}^1)^*(s_1^*) \otimes g_{n,X}^*(N_{Y/X})$$
$$\cong s_n^* \otimes \ldots \otimes s_1^* \otimes N_{Y/X}. \quad \square$$

In the case of a smooth curve $C \subset X$ we want to describe the embedding $F_n(C) \subset F_n(X)$ a little more precisely.

Remark 3.3.4. Let $C \subset X$ be a smooth curve. As $G_n(C)$ has rank 1 over $F_n(C)$, the projection $f_{n,C} : F_n(C) \longrightarrow F_{n-1}(C)$ is an isomorphism and so also $g_{n,C} : F_n(C) \longrightarrow C$ is an isomorphism. The embedding

$$j_{n,C} : F_{n-1}(C) \xrightarrow{f_{n,C}^{-1}} F_n(C) \hookrightarrow f_{n,X}^{-1}(F_{n-1}(C))$$

is defined by the sub line bundle $T_{F_{n-1}(C)}$ of $G_{n-1}(X) \subset T_{F_{n-1}(X)}$. Let $i_C : C \longrightarrow X$ be the embedding of C into X and $i_{n,C}$ the embedding

$$i_{n,C} : C \xrightarrow{g_{n,C}^{-1}} F_n(C) \hookrightarrow g_{n,X}^{-1}(C).$$

Then we obviously have

$$i_{n,C} = j_{n,C} \circ \ldots \circ j_{1,C} \circ i_C.$$

Remember that $X_{(n),c}^{[n]} \subset X^{[n]}$ parametrizes subschemes of the form $spec(\mathcal{O}_C/\mathbf{m}_{X,x}^n)$ for smooth locally closed curves $C \subset X$ and $x \in C$.

Lemma 3.3.5. *The map*

$$spec(\mathcal{O}_C/\mathbf{m}_{X,x}^{n+1}) \longmapsto i_{n,C}(x)$$

defines an open embedding

$$i_n : X_{(n+1),c}^{[n+1]} \longrightarrow F_n(X)$$

with image $F_n(X)_0$ (see definition 3.3.1).

Proof: We have to show that this map is well defined (i.e. does not depend on the choice of the smooth curve C) and defines an isomorphism. For this we introduce local coordinates on $X^{[n+1]}_{(n+1),c}$ and $F_n(X)_0$. Let $Z \in X^{[n+1]}_{(n+1),c}$. Let $(x_1, \dots, x_d)$ be local coordinates on $U \subset X$ such that

$$I_Z := (x_1^{n+1}, x_2, \dots, x_d)$$

is the ideal of Z. The subschemes $Z' \in X^{[n+1]}_{(n+1),c}$ near Z are defined by ideals

$$I_{Z'} := \left((x_1 - a_{1,0})^{n+1}, \left(x_i - \sum_{j=0}^{n} a_{i,j} x_1^j \right)_{i=2,\dots,d} \right).$$

So $a_{1,0}$ and the $a_{i,j}$ $(i = 2, \dots, d, \ j = 0, \dots, n)$ are local coordinates of $X^{[n+1]}_{(n+1),c}$ near Z.

We want to suppress the pullback in the notation. We write $x_i^0 := x_i$. Let $V \subset f_1^{-1}(U)$ be the open subset on which $dx_1^0|_{s_1} \neq 0$ holds. Then

$$x_i^1 := \frac{dx_i^0|_{s_1}}{dx_1^0|_{s_1}}$$

is regular for $i = 2, \dots, d$. x_1^0 and the dx_i^1 $(i = 2, \dots, d)$ form a basis of the relative differentials $\Omega_{F_1(X)/X}|_V$.

Let by induction x_1^0 and x_i^j, $(i = 2, \dots, d, \ j = 0, \dots, n)$ be local coordinates on $(g_n^1)^{-1}(V) \cap F_n(X)_0$ such that the dx_i^n $(i = 2, \dots, d)$ form a basis of $\Omega_{F_n(X)/F_{n-1}(X)}|_{(g_n^1)^{-1}(V) \cap F_n(X)_0}$. Then we have $dx_1^0|_{s_{n+1}} \neq 0$ on $(g_{n+1}^1)^{-1}(V) \cap F_{n+1}(X)_0$, and the functions

$$x_i^{n+1} := \frac{dx_i^n|_{s_{n+1}}}{dx_1^0|_{s_{n+1}}}$$

are regular on $(g_{n+1}^1)^{-1}(V) \cap F_{n+1}(X)_0$. x_1^0 and the x_i^j $(i = 2, \dots, d; \ j = 0, \dots, n+1)$ are local coordinates on $(g_{n+1}^1)^{-1}(V) \cap F_{n+1}(X)_0$. The dx_i^{n+1} $(i = 2, \dots, d)$ form a basis of $\Omega_{F_{n+1}(X)/F_n(X)}|_{(g_{n+1}^1)^{-1}(V) \cap F_{n+1}(X)_0}$. These coordinates have been introduced in the case of a surface in [Colley-Kennedy (1)].

Now let $C \subset U$ be a smooth locally closed curve such that x_1 is a local parameter on C. From the definitions we get that in our coordinates the map $i_{n,C}$ is given by

$$x \longmapsto \left(x_1(x), \left(\left(\frac{d}{d(x_1|_C)} \right)^j (x_i|_C)(x) \right)_{i=2,\dots,d; \ j=0,\dots,n} \right).$$

So the map

$$spec(\mathcal{O}_C / \mathbf{m}_{X,x}^{n+1}) \longmapsto i_{n,C}(x)$$

can be described in our local coordinates by

$$(a_{1,0}, (a_{i,j})_{i=2,\dots,d;\ j=0,\dots,n}) \mapsto (a_{1,0}, (b_{i,j})_{i=2,\dots,d;\ j=0,\dots,n})$$

where

$$b_{i,j} = \left(\frac{d}{dx}\right)^j \left(\sum_{k=0}^{n} a_{i,k} x^k\right)\Bigg|_{x=a_{1,0}}$$

$$= j! a_{i,j} + \sum_{k>j} \frac{k!}{(k-j)!} a_{i,k} a_{1,0}^{k-j}.$$

So it is well-defined and an isomorphism on $(g_n^1)^{-1}(V)$. As the inverse $i_{n,C}(x) \mapsto spec(\mathcal{O}_C/\mathbf{m}_{X,x}^{n+1})$ is well-defined and does not depend on the local coordinates, i_n is an isomorphism onto its image. We see that we can cover all of $F_n(X)_0$ by changing the local coordinates. As i_n is an isomorphism in all coordinate charts, its image is the whole of $F_n(X)_0$. □

So we see that $F_n(X)$ is a smooth compactification of $X_{(n+1),c}^{[n+1]}$.

Now we want to compute the number of nth order contacts of a smooth variety $X \subset \mathbf{P}_N$ with lines in $\mathbf{P}_N$.

Definition 3.3.6. Let $Al_{n+1}^{n+1}(\mathbf{P}_N) \subset \mathbf{P}_N^{[n+1]}$ be the closed subvariety

$$Al_{n+1}^{n+1}(\mathbf{P}_N) := \left\{ Z \in \mathbf{P}_N^{[n+1]} \ \middle| \ \begin{array}{c} Z \text{ is subscheme of a line,} \\ \text{and the suppport of } Z \text{ is one point} \end{array} \right\}$$

with the reduced induced structure.

Obviously $Al_{n+1}^{n+1}(\mathbf{P}_N)$ is a subvariety of $(\mathbf{P}_N)_{(n+1),c}^{[n+1]}$.

Now we want to describe $Al_{n+1}^{n+1}(\mathbf{P}_N)$. By definition it parametrizes subschemes of the form $spec(\mathcal{O}_l/\mathbf{m}_{X,x}^{n+1})$ for lines $l \subset \mathbf{P}_N$ and points $x \in l$. Let $A(N) \subset \mathbf{P}_N \times G(1,N)$ be the incidence variety

$$A(N) := \left\{ (x,l) \in \mathbf{P}_N \times G(1,N) \ \middle| \ x \in l \right\}.$$

Lemma 3.3.7. *Let $n \geq 1$. The application*

$$spec(\mathcal{O}_l/\mathbf{m}_{X,x}^{n+1}) \longmapsto (x,l) \in \mathbf{P}_n \times G(1,N)$$

gives an isomorphism

$$e_n : Al_{n+1}^{n+1}(\mathbf{P}_N) \longrightarrow A(N) \subset \mathbf{P}_N \times G(1,N).$$

Proof: Let $x_1, \ldots, x_N$ be the standard coordinates on $\mathbf{A}^N \subset \mathbf{P}_N$. Let $Z \in Al_{n+1}^{n+1}(\mathbf{P}_N)$. We can assume that $Z \subset \mathbf{A}^N$, and that the ideal of Z is of the form $I_Z := (x_1^{n+1}, x_2, \ldots, x_N)$. A subscheme $Z' \in Al_{n+1}^{n+1}(\mathbf{P}_N)$ near Z has an ideal of the form

$$I_{Z'} := \left((x_1 - a_{1,0})^{n+1}, x_2 - a_{2,1}x_1 - a_{2,0}, \ldots, x_N - a_{N,0}x_1 - a_{N,0}\right),$$

and $a_{1,0}$ and the $a_{i,0}, a_{i,1}, (i = 2, \ldots, N)$ are local coordinates on $Al_{n+1}^{n+1}(\mathbf{P}_N)$ near Z.

Let l be the line defined by $(x_2, \ldots, x_N)$. A line near l is given by

$$(x_2 - a_{2,1}x_1 - a_{2,0}, \ldots, x_N - a_{N,0}x_1 - a_{N,0}),$$

and the $a_{i,0}, a_{i,1}, (i = 2, \ldots, N)$ are local coordinates on $G(1, N)$ near l. So the application

$$e_n : Al_{n+1}^{n+1}(\mathbf{P}_N) \longrightarrow \mathbf{P}_N \times G(1, N)$$

is given in our local coordinates by

$$(a_{1,0}, a_{2,0}, \ldots, a_{N,0}, a_{2,1} \ldots, a_{N,1}) \longmapsto$$
$$\left((a_{1,0}, a_{2,0}, \ldots, a_{N,0}), (a_{2,0}, \ldots, a_{N,0}, a_{2,1} \ldots, a_{N,1})\right),$$

and this defines an isomorphism with the subvariety

$$A(N) \subset \mathbf{P}_N \times G(1, N) \qquad \square$$

Remark 3.3.8. Let $X \subset \mathbf{P}_N$ be a smooth subvariety of codimension r. From the definitions we can see that the intersection points of $X_{(n+1),c}^{[n+1]}$ and $Al_{n+1}^{n+1}(\mathbf{P}_n)$ in $(\mathbf{P}_N)_{(n+1),c}^{[n+1]}$ correspond exactly to the nth order contacts of X with lines in $\mathbf{P}_N$. More precisely we have: the image $e_n(X_{(n+1),c}^{[n+1]} \cap Al_{n+1}^{n+1}(\mathbf{P}_N))$ is

$$\left\{ (x, l) \in X \times G(1, N) \;\middle|\; l \text{ has } n^{th} \text{ order contact with } X \text{ at } x \right\}.$$

Now we want to describe the incidence variety $A(N) \subset \mathbf{P}_N \times G(1, N)$ more precisely. We have the projections

$$A(N)$$
$$\swarrow {\scriptstyle p_1} \qquad \qquad \searrow {\scriptstyle p_2}$$
$$\mathbf{P}_N \qquad \qquad \qquad G(1, N)$$

Remark 3.3.9. Let $\mathcal{O}_{\mathbf{P}_N}(-1)$ be the tautological line bundle on $\mathbf{P}_N$ and $T_2 := T_{2,N+1}$ the tautological subbundle on $G(1,N) = Gr(2, N+1)$. Let $Q_1 := Q_{N,N+1} := \mathcal{O}_{\mathbf{P}_N}^{\oplus N+1}/\mathcal{O}_{\mathbf{P}_N}(-1)$. Then we can see easily that p_1 and p_2 can be described as the natural projections

$$\mathbf{P}(Q_1) = A(N) = \mathbf{P}(T_2)$$

$$\swarrow p_1 \qquad\qquad\qquad \searrow p_2$$

$$\mathbf{P}_N \qquad\qquad\qquad\qquad\qquad G(1,N)$$

(see also [Fulton (1)] Ex 14.7.12). We put

$$\tilde{H} := p_1^*(\mathcal{O}_{\mathbf{P}_N}(1)),$$
$$\tilde{P} := \mathcal{O}_{\mathbf{P}(Q_1)}(1),$$
$$H := c_1(\tilde{H}),$$
$$P := c_1(\tilde{P}).$$

Then we can see easily that $\tilde{H} = \mathcal{O}_{\mathbf{P}(T_2)}(1)$, and $\tilde{P}^*$ is the universal quotient bundle

$$\tilde{P}^* = Q_{1,T_2} = p_2^*(T_2)/\mathcal{O}_{\mathbf{P}(T_2)}(-1) = p_2^*(T_2)/\tilde{H}^*.$$

We have

$$p_1^*(c(Q_1)) = 1 + H + H^2 + \ldots + H^N,$$

and so

$$A^*(A(N)) = \frac{\mathbb{Z}[H,P]}{\left(H^{N+1}, \ \sum_{i=0}^{N} H^i P^{N-i}\right)}. \qquad (*)$$

Let $Q_2 := Q_{N-1,N+1} := (\mathcal{O}_{G(1,N)}^{\oplus N+1}/T_2)$ be the universal quotient bundle on $G(1,N)$. $p_2^*(c_k(Q_2))$ is the pullback of a Schubert cycle

$$\left\{(x,l) \in A(N) \ \middle| \ \begin{array}{c} l \text{ intersects a fixed} \\ (k+1)\text{-codimensional linear subspace} \end{array} \right\}.$$

$\tilde{H}^* = \mathcal{O}_{\mathbf{P}(T_2)}(-1)$ and $\tilde{P}^* = Q_{1,T_2}$ imply $p_2^*(Q_2) = p_1^*(Q_1)/\tilde{P}^*$ and so

$$p_2^*(c_k(Q_2)) = \sum_{j=0}^{k} H^j P^{k-j}.$$

The relative tangent bundle is

$$T_{A(N)/G(1,N)} = \mathcal{O}_{\mathbf{P}(T_2)}(1) \otimes (p_2^*(T_2)/\mathcal{O}_{\mathbf{P}(T_2)}(-1)) = \tilde{H} \otimes \tilde{P}^*.$$

Now we want to describe the restriction

$$i_n|_{A_{N+1}^{N+1}(\mathbf{P}_N)} : A_{N+1}^{N+1}(\mathbf{P}_N) \longrightarrow F_n(\mathbf{P}_N).$$

For this we give embeddings $\alpha_n : A(N) \longrightarrow F_n(\mathbf{P}_N)$ with $i_n|_{A_{N+1}^{N+1}(\mathbf{P}_N)} = \alpha_n \circ e_n$.

Definition 3.3.10. We want to define $\alpha_n : A(N) \longrightarrow F_n(\mathbf{P}_N)$ for $n \geq 1$ inductively. We have

$$T_{\mathbf{P}_N} = \mathcal{O}_{\mathbf{P}_N}(1) \otimes Q_1.$$

So there is a natural isomorphism $\alpha_1 : A(N) = \mathbf{P}(Q_1) \longrightarrow \mathbf{P}(T_{\mathbf{P}_N})$ with

$$\begin{aligned}
\alpha_1^*(s_1) &= \alpha_1^*(\mathcal{O}_{\mathbf{P}(T_{\mathbf{P}_N})}(-1)) \\
&= \tilde{H} \otimes \tilde{P}^* \\
&= T_{A(N)/G(1,N)}.
\end{aligned}$$

We put $A_1 := F_1(\mathbf{P}_N)$. A_1 is mapped to $G(1,N)$ by $p_2 \circ \alpha_1^{-1}$. $T_{A_1/G(1,N)}$ is a sub line bundle of $T_{F_1(\mathbf{P}_N)}$, and the diagram

$$\begin{array}{ccc}
T_{A_1/G(1,N)} & \hookrightarrow & T_{F_1(\mathbf{P}_N)} \\
\downarrow & & \downarrow {\scriptstyle df_1} \\
s_1 & \hookrightarrow & f_1^*(T_{\mathbf{P}_N})
\end{array}$$

commutes. So $T_{A_1/G(1,N)}$ is a sub line bundle of $G_1(\mathbf{P}_N) \subset T_{F_1(\mathbf{P}_N)}$.

We assume by induction that $\alpha_n : A(N) \longrightarrow F_n(\mathbf{P}_N)$ is an embedding. Let $A_n \subset F_n(\mathbf{P}_N)$ be its image. A_n is mapped to $G(1,N)$ by $p_2 \circ \alpha_n^{-1}$. We also assume that $T_{A_n/G(1,N)}$ is a sub line bundle of $G_n(\mathbf{P}_N)|_{A_n}$. Let

$$\beta_{n+1} : A_n \longrightarrow f_{n+1}^{-1}(A_n) \subset F_{n+1}(\mathbf{P}_N)$$

be the embedding defined by the sub line bundle $T_{A_n/G(1,N)}$ of $G_n(\mathbf{P}_N)|_{A_n}$. Let $A_{n+1} \subset F_{n+1}(\mathbf{P}_N)$ be the image of β_{n+1}. Then $T_{A_{n+1}/G(1,N)}$ is a sub line bundle of $T_{F_{n+1}(\mathbf{P}_N)}|_{A_{n+1}}$, and the diagram

$$\begin{array}{ccc}
T_{A_{N+1}/G(1,N)} & \hookrightarrow & T_{F_{N+1}(\mathbf{P}_N)} \\
\downarrow & & \downarrow {\scriptstyle df_{N+1}} \\
s_{N+1} & \hookrightarrow & f_{N+1}^*(T_{\mathbf{P}_N})
\end{array}$$

commutes. So $T_{A_{N+1}/G(1,N)}$ is a sub line bundle of

$$G_{N+1}(\mathbf{P}_N)|_{A_{n+1}} \subset T_{F_{n+1}(\mathbf{P}_N)}|_{A_{n+1}}.$$

We put $\alpha_{n+1} := \beta_{n+1} \circ \alpha_n$. This is a closed embedding. We get inductively for all n:

$$\alpha_n^*(s_n) = T_{A(N)/G(1,N)} = \tilde{H} \otimes \tilde{P}^*.$$

Lemma 3.3.11. $i_n|_{A_{N+1}^{N+1}(\mathbf{P}_N)} = \alpha_n \circ e_n.$

Proof: We only have to show that

$$i_n|_{l_{(N+1)}^{[N+1]}} = \alpha_n \circ e_n|_{l_{(N+1)}^{[N+1]}}$$

holds for every line. Here $l_{(n+1)}^{[n+1]}$ is the closed subvariety of $l^{[n+1]}$ parametrizing subschemes of length $n + 1$ which are concentrated in a point of l. The projection $p : l_{(N+1)}^{[N+1]} \longrightarrow l$ mapping such a subscheme to its support is an isomorphism and $\epsilon := e_n|_{l_{(N+1)}^{[N+1]}} \circ p^{-1}$ is the map

$$\epsilon : l \longrightarrow A(N); \quad x \longmapsto (x, l).$$

So we have to show that $\alpha_n \circ \epsilon = i_{n,l}$ holds for the embedding $i_{n,l} : l \longrightarrow g_{n,\mathbf{P}_N}^{-1}(l) \subset F_n(X)$.

By definition 3.3.10 $\alpha_1 : A(N) \longrightarrow \mathbf{P}(T_{\mathbf{P}_N})$ is defined by the sub line bundle $T_{A(N)/G(1,N)} \subset T_{\mathbf{P}_N}$. So the sub line bundle $T_l \subset T_{\mathbf{P}_N}|_l$ defines the embedding

$$\alpha_1 \circ \epsilon : l \longrightarrow f_1^{-1}(l) \subset F_1(\mathbf{P}_N).$$

By remark 3.3.4 this also defines $i_{1,l}$. By induction we assume that $\alpha_n \circ \epsilon = i_{n,l}$. In particular we have

$$(\alpha_n \circ \epsilon)(l) = F_n(l) \subset g_{n,\mathbf{P}_N}^{-1}(l).$$

The embedding $\beta_{n+1} : A_n \longrightarrow f_{n+1}^{-1}(A_n)$ is given by the sub line bundle $T_{A_n/G(1,N)} \subset G_n(\mathbf{P}_N)|_{A_n}$, i.e.

$$\beta_{n+1}|_{F_n(l)} : F_n(l) \longrightarrow f_{n+1}^{-1}(F_n(l))$$

is given by $T_{F_n(l)} \subset G_n(\mathbf{P}_N)|_{F_n(l)}$. By remark 3.3.4 this also defines the embedding

$$j_{n+1,l} : F_n(l) \longrightarrow f_{n+1}^{-1}(F_n(l)).$$

So we have $\beta_{n+1}|_{F_n(l)} = j_{n+1,l}$, and thus by remark 3.3.4

$$\alpha_{n+1} \circ \epsilon = j_{n+1,l} \circ i_{n,l} = i_{n+1,l}. \qquad \square$$

Now we can show a general formula for the numbers of higher order contacts of X with lines in $\mathbf{P}_N$.

Definition 3.3.12. Let $X \subset \mathbf{P}_N$ be a smooth projective subvariety. Let $p : Al_{n+1}^{n+1}(\mathbf{P}_N) \longrightarrow \mathbf{P}_N$ be the projection. We put

$$Al_{n+1,X} := p^{-1}(X).$$

Let $p_X : Al_{n+1,X} \longrightarrow X$ be the restriction of p. Let

$$k_{n,X} : Al_{n+1,X} \longrightarrow g_{n,\mathbf{P}_N}^{-1}(X)$$

be the restriction of the embedding $i_n : (\mathbf{P}_N)_{(n+1),c}^{[n+1]} \longrightarrow F_n(\mathbf{P}_N)$ to $Al_{n+1,X}$. Let $[F_n(X)]$ be the class of $F_n(X)$ in $A^*(g_{n,\mathbf{P}_N}^{-1}(X))$. The class of nth order contacts of X with lines $\mathbf{P}_N$ is defined as

$$K_n(X) := (p_X)_*(k_{n,X}^*([F_n(X)])) \in A^*(X).$$

The class of nth order contacts of X with lines in $\mathbf{P}_N$ which intersect a general linear subvariety of dimension $l+1$ is

$$K_{n,l}(X) := (p_X)_*(k_{n,X}^*([F_n(X)]) \cdot e_n^*(p_2^*(c_l(Q_2)))) \in A^*(X).$$

For a closed subvariety $X \subset \mathbf{P}_N$ we put $A_X := p_1^{-1}(X) \subset A(N)$. Let $q_1 : A_X \longrightarrow X$ and $q_2 : A_X \longrightarrow G(1, N)$ be the projections. Let

$$\alpha_{n,X} : A_X \longrightarrow g_{n,\mathbf{P}_N}^{-1}(X)$$

be the restriction of $\alpha_n : A(N) \longrightarrow F_n(\mathbf{P}_N)$.

Remark 3.3.13. By lemma 3.3.11 we get

$$K_n(X) = (q_1)_*(\alpha_{n,X}^*([F_n(X)])),$$
$$K_{n,l}(X) = (q_1)_*(\alpha_{n,X}^*([F_n(X)]) \cdot q_2^*(c_l(Q_2))).$$

Remark 3.3.14. Let $h \subset \mathbf{P}_N$ be a general linear subspace of codimension $l+1$ and $W(h) \subset G(1, N)$ the set of lines intersecting h. By remark 3.3.8, lemma 3.3.11, definition 3.3.12 and remark 3.3.13 we have:

$$\alpha_{n,X}^{-1}(F_n(X)) = \left\{ (x,l) \in A_X \; \middle| \; \begin{array}{c} l \text{ has } n^{th} \text{ order contact} \\ \text{with } X \text{ at } x \end{array} \right\},$$
$$q_1(\alpha_{n,X}^{-1}(F_n(X))) = \left\{ x \in X \; \middle| \; \begin{array}{c} \text{there is a line } l \text{ with which} \\ X \text{ has } n^{th} \text{ order contact at } x \end{array} \right\}$$

and

$$q_1(\alpha_{n,X}^{-1}(F_n(X)) \cap q_2^{-1}(W(h)))$$

$$= \left\{ x \in X \;\middle|\; \begin{array}{c} \text{there is a line } l \\ \text{intersecting } h \text{ and having } n^{th} \text{ order} \\ \text{contact with } X \text{ at } x \end{array} \right\}.$$

Let W be an irreducible component of $q_1(\alpha_{n,X}^{-1}(F_n(X)) \cap q_2^{-1}(W(h)))$. The multiplicity of W in $(q_1)_*(\alpha_{n,X}^{-1}(F_n(X)) \cap q_2^{-1}(W(h)))$ is the degree of $q_1|_{\alpha_{n,X}^{-1}(F_n(X)) \cap q_2^{-1}(W(h))}$ over W (or 0 if this degree is infinite), i.e. the number of lines intersecting h having n^{th} order contact with X in a general point $x \in W$ counted with multiplicity. In particular we have: let $Y \subset X$ be a closed subvariety of dimension d where $d = l + nr - N + 1$ so that there are only finitely many n^{th} order contacts of X with lines intersecting h in points of Y. Then the number of these contacts counted with multiplicities is the intersection number $K_{n,l}(X) \cdot [Y]$.

Theorem 3.3.15. *Let n be a positive integer. Let $X \subset \mathbf{P}_N$ be a smooth projective variety of codimension r, let $N_{X/\mathbf{P}_N}$ be the normal bundle of X in $\mathbf{P}_N$ and H the class of a hyperplane section. Let $0 \le l \le N$ and $d := l + nr - N + 1$. We assume $0 \le d \le N - r$. Then we have:*

$$K_{n,l}(X) = \sum_{k=0}^{d} \left(\sum_{s=max(0,k-l)}^{k} (-1)^s \binom{N+k-l}{s} \right) \cdot$$

$$\sum_{i_1+\ldots+i_n=d-k} \left(\prod_{j=1}^{n} j^{r-i_j} \right) H^k \left(\prod_{j=1}^{n} c_{i_j}(N_{X/\mathbf{P}_N}) \right) \in A^d(X).$$

In particular we have in the case $l = 0$:

$$K_n(X) = \sum_{k=0}^{d} (-1)^k \binom{N+k}{k} \sum_{i_1+\ldots+i_n=d-k} \left(\prod_{j=1}^{n} j^{r-i_j} \right) H^k \left(\prod_{j=1}^{n} c_{i_j}(N_{X/\mathbf{P}_N}) \right).$$

Let $Y \subset X$ be a closed subvariety of dimension d and $[Y] \in A^{N-r-d}(X)$ its class. Let $h \subset \mathbf{P}_N$ be a general $(l+1)$-codimensional linear subvariety. If there are only finitely many n^{th} order contacts of X with lines intersecting h in points of Y, then the number of these contacts counted with multiplicities is

$$\sum_{k=0}^{d} \left(\sum_{s=max(0,k-l)}^{k} (-1)^s \binom{N+k-l}{s} \right) \cdot$$

$$\sum_{i_1+\ldots+i_n=d-k} \left(\prod_{j=1}^{n} j^{r-i_j} \right) H^k \left(\prod_{j=1}^{n} c_{i_j}(N_{X/\mathbf{P}_N}) \right) \cdot [Y].$$

If in particular $l = 0$ and $d = N - r = dim(X)$, and so $2N - 1 = (n+1)r$, then the number of nth order contacts of X with lines in $\mathbf{P}_N$ counted with multiplicities is

$$\sum_{k=0}^{N-r} (-1)^k \binom{N+k}{k} \sum_{i_1+\ldots+i_n=N-r-k} \left(\prod_{j=1}^{n} j^{r-i_j} \right) H^k \left(\prod_{j=1}^{n} c_{i_j}(N_{X/\mathbf{P}_N}) \right).$$

We first show the following lemma:

Lemma 3.3.16.

(1)
$$(q_1)_*(P^{N-1+k}) = \begin{cases} 0, & k < 0; \\ 1, & k = 0; \\ -H, & k = 1; \\ 0, & otherwise. \end{cases}$$

(2)
$$(q_1)_* \left((P - H)^{N-1+k-l} \sum_{t=0}^{l} H^t P^{l-t} \right)$$

$$= \sum_{s=max(0,k-l)}^{k} (-1)^s \binom{N+k-l}{s} H^k.$$

Proof:

(1) By remark 3.3.9(∗) we get

$$A^*(A_X) = \frac{A^*(X)[P]}{\left(\displaystyle\sum_{i=0}^{N} H^i P^{N-i} \right)}. \tag{∗∗}$$

The result is clear for $k \leq 0$ and $k > N$. By (∗∗) and the projection formula we get

$$(q_1)_*(P^N) = (q_1)_*(-HP^{N-1}) = -H.$$

Now let $N \geq k \geq 2$, and assume the result holds for $k - 1$. Then we get by (∗∗) and the projection formula

$$(q_1)_*(P^{N-1+k}) = -\sum_{s=1}^{N} H^s (q_1)_*(P^{N-1+k-s}).$$

By induction and the above this is $-H^k + H^k = 0$.

(2) By (1) we have

$$(q_1)_* \left((P - H)^{N-1+k-l} \sum_{t=0}^{l} H^t P^{l-t} \right)$$

$$= (q_1)_* \left((P - H)^{N-1+k-l} \sum_{s=k-l}^{k} H^{k-s} P^{l-k+s} \right)$$

$$= H^k \sum_{s=max(0,k-l)}^{k} (-1)^s \left(\binom{N-1+k-l}{s} + \binom{N-1+k-l}{s-1} \right)$$

$$= H^k \sum_{s=max(0,k-l)}^{k} (-1)^s \binom{N+k-l}{s}. \qquad \square$$

Proof of theorem 3.3.15: We only have to show the formula for $K_{n,l}(X)$. By lemma 3.3.3 and definition 3.3.10 we have

$$\alpha_{n,X}^*([F_n(X)]) = c_r(q_1^*(N_{X/\mathbf{P}_N}) \otimes \alpha_{1,X}^*(s_1^*)) \cdot \ldots$$
$$\cdot c_r(q_1^*(N_{X/\mathbf{P}_N}) \otimes \alpha_{1,X}^*(s_1^*) \otimes \ldots \otimes \alpha_{n,X}^*(s_n^*)))$$

$$= \prod_{j=1}^{n} c_r \left(q_1^*(N_{X/\mathbf{P}_N}) \otimes (\tilde{P} \otimes \tilde{H}^*)^{\otimes j} \right)$$

$$= \prod_{j=1}^{n} \left(\sum_{i=1}^{r} j^{r-i} q_1^*(c_i(N_{X/\mathbf{P}_N}))(P - H)^{r-i} \right)$$

So we get by the projection formula and remark 3.3.9:

$$K_{n,l}(X)$$
$$= (q_1)_*(\alpha_{n,X}^*([F_n(X)]) \cdot q_2^*(c_l(Q_2)))$$
$$= (q_1)_* \left(q_2^*(c_l(Q_2)) \prod_{j=1}^{n} \left(\sum_{i=1}^{r} j^{r-i} q_1^*(c_i(N_{X/\mathbf{P}_N}))(P - H)^{r-i} \right) \right)$$
$$= \sum_{k=0}^{d} \sum_{i_1+\ldots+i_n=d-k} (q_1)_* \left(\sum_{t=0}^{l} H^t P^{l-t}(P - H)^{nr-d+k} \right) \prod_{j=1}^{r} j^{r-i_j} c_{i_j}(N_{X/\mathbf{P}_N}).$$

By the definitions we have $nr - d + k = N - 1 + k - l$. The result now follows by lemma 3.3.16. $\square$

So we have found formulas for the contacts of $X \subset \mathbf{P}_N$ with lines in $\mathbf{P}_N$ as linear combinations of monomials in H and the Chern classes $c_i(N_{X/\mathbf{P}_N})$. Using the formula

$$c(N_{X/\mathbf{P}_N}) = (1 + H)^{N+1}/c(T_X)$$

we can replace the Chern classes of $N_{X/\mathbf{P}_N}$ by those of X if we want. The result will however be more complicated this way. It is easy to check that the formulas after proposition 3.2.19 can be obtained as special cases.

Now we want to show that more generally the class in $A^*(X)$ of the locus where a smooth projective variety $X \subset \mathbf{P}_N$ has n^{th} order contact with a given family of curves is properly interpreted a linear combination of monomials in H and the Chern classes of X. This linear combination will depend on the family C_T. We will not treat here the much more difficult question how to determine this linear combination for a given family C_T. The argument is similar to that at the end of section 3.2. First we will generalize the Semple bundles to a relative situation.

Definition 3.3.17. Let T be an irreducible algebraic variety. Let $X \longrightarrow T$ be a smooth morphism of relative dimension d. We will inductively define varieties $F_n(X/T)$ and vector bundles $G_n(X/T)$ on $F_n(X/T)$. Let $F_0(X/T) :=$ X, $G_0(X/T) := T_{X/T}$. By induction assume that $F_0(X/T), \ldots, F_{n-1}(X/T)$ are already defined. Assume that $G_{n-1}(X/T)$ is a subbundle of rank d of $T_{F_{n-1}(X)/T}$. Then we put

$$F_n(X/T) := \mathbf{P}(G_{n-1}(X/T)).$$

Let

$$f_{n,X/T} : \mathbf{P}(G_{n-1}(X/T)) \longrightarrow F_{n-1}(X/T)$$

be the projection. Let s_n be the tautological subbundle of $f^*_{n,X/T}(G_{n-1}(X/T))$. Now we define the subbundle $G_n(X)$ of $T_{F_n(X)/T}$ by the diagram

$$
\begin{array}{ccccccccc}
0 & \to & T_{F_n(X/T)/F_{n-1}(X/T)} & \to & T_{F_n(X/T)/T} & \to & f^*_{n,X/T}T_{F_{n-1}(X/T)/T} & \to & 0 \\
 & & \| & & \uparrow & \square & \uparrow & & \\
0 & \to & T_{F_n(X/T)/F_{n-1}(X/T)} & \to & G_n(X/T) & \to & s_n & \to & 0.
\end{array}
$$

If $Y \subset X$ is a (locally) closed subvariety such that the restriction to Y of the projection $X \longrightarrow T$ is a smooth morphism of relative dimension m, then we see in a similar way as in the proof of lemma 3.3.3 that $F_n(Y/T)$ is a (locally) closed subvariety of $F_n(X/T)$.

Definition 3.3.18. Let T be a smooth projective variety of dimension $m - 1$ and $C_T \subset \mathbf{P}_N \times T$ a flat family of curves, i.e. we have the projections

$$C_T$$

$$\swarrow p_1 \qquad\qquad \searrow p_2$$

$$\mathbf{P}_N \qquad\qquad\qquad T,$$

p_2 is flat and for all $t \in T$ the fibre $C_t = p_2^{-1}(t)$ is a curve. We assume in addition that there is a dense open subset $C_{T,0} \subset C_T$ such that the restriction $C_{T,0} \longrightarrow T$ is a smooth morphism.

Then $F_n(C_{T,0}/T)$ is a locally closed subvariety of

$$F_n((\mathbf{P}_N \times T)/T) = F_n(\mathbf{P}_N) \times T.$$

Let $\bar{F}_n(C_T)$ be the closure of $F_n(C_{T,0}/T)$ in $F_n(\mathbf{P}_N) \times T$ and $[\bar{F}_n(C_T)]$ its class in $A^{(N-1)(n+1)}(F_n(\mathbf{P}_N) \times T)$. Let

$$p : F_n(\mathbf{P}_N) \times T \longrightarrow F_n(\mathbf{P}_N)$$

be the projection. We put

$$K_n(C_T) := p_*([\bar{F}_n(C_T)]) \in A^r(F_n(\mathbf{P}_N)),$$

where $r := N + (N-1)n - m$. Let $X \subset \mathbf{P}_N$ be a smooth projective variety of dimension d. Let $i_{n,X} : F_n(X) \longrightarrow F_n(\mathbf{P}_N)$ be the embedding. We put

$$K_n(X, C_T) := (g_{n,X})_*(i_{n,X}^*(K_n(C_T)) \in A^e(X),$$

where $e = N + (N-d)n - m$.

Remark 3.3.19. $K_n(X, C_T)$ is a candidate for the class of the locus where X and curves in the family C_T have n^{th} order contact. We have for example

$$g_{n,X}(p(F_n(C_{T,0}/T) \cap (F_n(X) \times T)))$$
$$= \left\{ x \in X \;\middle|\; \begin{array}{l} \text{there is a } t \in T \text{ such that } x \text{ is a smooth point of} \\ C_t \text{ and } C_t \text{ has } n^{th} \text{ order contact with } X \text{ in } x \end{array} \right\}.$$

Assume in particular $e = d$, i.e. $m = (n+1)(N-d)$, and assume the subset

$$F_n(C_{T,0}/T) \cap (F_n(X) \times T) \subset F_n(\mathbf{P}_N) \times T$$

to be finite and to coincide with $\bar{F}_n(C_T) \cap (F_n(X) \times T)$. Then the number of nth order contacts of X with curves in the family C_T counted with multiplicities is $K_n(X, C_T)$.

Proposition 3.3.20. *Let* $n, d \in \mathbb{N}$. *Let* C_T *be a family of curves satisfying the conditions of definition 3.3.18 with* $\dim(C_T) = m$. *Assume* $e = N + (N-d)n - m$ *and* $0 \le e \le d$. *For all partitions*

$$(\alpha) = (1^{\alpha_1}, 2^{\alpha_2}, \ldots)$$

of numbers $s \leq e$ there are integers n_α such that for all closed subvarieties $X \subset \mathbf{P}_N$ of dimension d

$$K_n(X, C_T) = \sum_{s=0}^{e} \sum_{\alpha \in P(s)} n_\alpha H^{e-s} c_1(X)^{\alpha_1} \ldots c_e(X)^{\alpha_e}.$$

Proof: We show more generally that for any $W \in A^{e+n(d-1)}(F_n(\mathbf{P}_N))$ and for all partitions α of $s \leq e$ there are integers n_α, satisfying

$$(g_{n,X})_*(i_{n,X}^*(W)) = \sum_{s=0}^{e} \sum_{\alpha \in P(s)} n_\alpha H^{e-s} c_1(X)^{\alpha_1} \ldots c_d(X)^{\alpha_d}.$$

As $A^*(F_{n-1}(\mathbf{P}_N))$ is generated by $H, P_1 := c_1(s_1^*), \ldots, P_{n-1} := c_1(s_{n-1}^*)$, it is enough to prove the result for monomials in $H, P_1, \ldots, P_n$. We will now suppress $i_{n,X}^*$ in the notation and write g_n instead of $g_{n,X}$. Let $M = M_0 P_1^{l_1} \ldots P_n^{l_n}$ be a monomial. Here M_0 is a monomial in $H, c_1(X), \ldots, c_d(X)$. If $l_i = d - 1$ for all $i = 1, \ldots, n$, then we have $(g_n)_*(M) = M_0$. Otherwise let j_0 be the largest j such that $l_j \neq d - 1$. By proposition 3.3.2 we see that $(g_n)_*(M) = 0$ if $l_{j_0} < d - 1$. So let $l_{j_0} \geq d$. By proposition 3.3.2 we can express M as a linear combination with $\mathbb{Z}$-coefficients of monomials $N = N_0 P_1^{m_1} \ldots P_n^{m_n}$, where N_0 is a monomial in $H, c_1(X), \ldots, c_d(X)$, and we have $m_j = l_j$ for $j > j_0$ and $m_{j_0} < l_{j_0}$. So the result follows by induction. $\square$

4. The Chow ring of relative Hilbert schemes of projective bundles

In this chapter we treat the Chow rings of relative Hilbert schemes of projectivizations of vector bundles over smooth projective varieties. In section 4.1 we will first construct embeddings of relative Hilbert schemes into Grassmannian bundles and study them. The case of the relative Hilbert scheme of a $\mathbf{P}_1$-bundle over a smooth variety is studied in more detail. From this we get the Chow ring of the variety $Al^n(\mathbf{P}_d)$ parametrizing subschemes of length n of $\mathbf{P}_d$ which lie on a line in $\mathbf{P}_d$. This variety has been used in [Le Barz (1),(2),(3),(4),(5),(8)] to obtain enumerative formulas for multisecants of curves and surfaces.

In section 4.2 we compute the Chow ring of the variety $\widetilde{\mathrm{Hilb}}^3(\mathbf{P}_2)$ parametrizing triangles in $\mathbf{P}_2$ with a marked side. This variety has been used in [Elencwajg-Le Barz (2),(3)] to compute the Chow ring of $\mathbf{P}_2^{[3]}$. The Chow ring of $\widetilde{\mathrm{Hilb}}^3(\mathbf{P}_2)$ has a much simpler structure than that of $\mathbf{P}_2^{[3]}$.

In section 4.3 we generalize this result to a relative situation. We compute the Chow ring of the variety $\widetilde{\mathrm{Hilb}}^3(\mathbf{P}(E)/X)$ parametrizing triangles with a marked side in the fibres of the projectivization $\mathbf{P}(E)$ of a vector bundle E. We also consider the variety $\widehat{H}^3(\mathbf{P}(E)/X)$ of complete triangles in the fibres of $\mathbf{P}(E)$, which has been studied in [Collino-Fulton (1)]. We pull back the classes in the Chow ring $A^*(\widetilde{\mathrm{Hilb}}^3(\mathbf{P}(E)/X)$ to $\widehat{H}^3(\mathbf{P}(E)/X)$ to find some of the relations. The most important case of our result is the variety $\widetilde{\mathrm{Cop}}^3(\mathbf{P}_d)$, parametrizing triangles with a marked side in $\mathbf{P}_d$ together with a plane containing them.

In section 4.4 we finally treat the relative Hilbert scheme $\mathrm{Hilb}^3(\mathbf{P}(E)/X)$ of subschemes of length 3 in the fibres of $\mathbf{P}(E)$. Analogously to [Elencwajg-Le Barz (3)] in the case of $\mathbf{P}_2$ we define a system of generators for the Chow ring of $\mathrm{Hilb}^3(\mathbf{P}(E)/X)$ as $A^*(X)$-algebra. By pulling these classes back to $\widetilde{\mathrm{Hilb}}^3(\mathbf{P}(E)/X)$ we determine their relations. To carry out the computations we have however to make use of a computer. The result is also quite complicated. The most important special case is again that of the variety $\mathrm{Cop}^3(\mathbf{P}_d)$, parametrizing pairs consisting of a subscheme of length 3 of $\mathbf{P}_d$ and a plane containing it. It can be obtained by blowing up $\mathbf{P}_d^{[3]}$ along $Al^3(\mathbf{P}_d)$. The Betti numbers of this variety have been determined in [Rosselló (1)]. In the case $d = 3$ it has been used in [Rosselló (2)] to determine the Chow ring of $\mathbf{P}_3^{[3]}$. In a recent joint work with Fantechi [Fantechi-Göttsche (1)] we have computed the cohomology ring $H^*(X^{[3]}, \mathbf{Q})$, for X an arbitrary smooth projective variety, by using an entirely different method.

4.1. n-very ampleness, embeddings of the Hilbert scheme and the structure of $Al^n(\mathbf{P}(E))$

Let X be a projective scheme over an algebraically closed field k. In [Beltrametti-Sommese (1)] the following definition was made:

Definition 4.1.1. Let $\mathcal{L}$ be an invertible sheaf on X. For every subscheme $Z \subset X$ we study the restriction map

$$r_{Z,\mathcal{L}} : H^0(X, \mathcal{L}) \longrightarrow H^0(X, \mathcal{L} \otimes \mathcal{O}_Z).$$

$\mathcal{L}$ is called n-very ample if $r_{Z,\mathcal{L}}$ is onto for every 0-dimensional subscheme $Z \subset X$ of length $len(Z) \leq n+1$.

Remark 4.1.2.

(1) We see that an invertible sheaf $\mathcal{L}$ is 0-very ample if and only if it is spanned by global sections and 1-very ample if and only if it is very ample.

(2) Let $\mathcal{L}$ be an $(n-1)$-very ample invertible sheaf on X. Then we can associate to each subscheme Z of length n on X the quotient

$$H^0(X, \mathcal{O}_Z \otimes \mathcal{L}) = H^0(X, \mathcal{L})/ker(r_{Z,\mathcal{L}})$$

of dimension n. This defines a morphism .

$$\phi_{\mathcal{L},n} : X^{[n]} \longrightarrow \mathrm{Grass}(n, H^0(X, \mathcal{L})).$$

It is clear from the definition that an n-very ample invertible sheaf is also m-very ample for every $m \leq n$. In [Beltrametti-Sommese (1)] only the case of a smooth surface S is considered. In this case they show that $\phi_{\mathcal{L},n}$ is injective if $\mathcal{L}$ is n-very ample and a closed embedding if $\mathcal{L}$ is $3n$-very ample. In the appendix [Göttsche (3)] of [Beltrametti-Sommese (1)] the corresponding very ample invertible sheaf on $S^{[n]}$ is identified. In [Catanese-Göttsche (1)] this result is sharpened and generalized to a general projective variety X. The main result is:

Theorem 4.1.3. [Catanese-Göttsche (1)] *Let X be a projective scheme over an algebraically closed field k and $\mathcal{L}$ an $(n-1)$-very ample invertible sheaf on X. The morphism*

$$\phi_{\mathcal{L},n} : X^{[n]} \longrightarrow \mathrm{Grass}(n, H^0(X, \mathcal{L}))$$

is an embedding if and only if $\mathcal{L}$ is n-very ample.

Now we want to generalize this result to a relative situation. Let T be a reduced projective variety and X a projective scheme over T. Let $\pi : X \longrightarrow T$ be the projection.

Definition 4.1.4. Let $\mathcal{L}$ be an invertible sheaf on X for which also $\pi_*(\mathcal{L})$ is locally free. For all $n \in \mathbb{N}$ let $\pi_n : \text{Hilb}^n(X/T) \longrightarrow T$ be the projection. Let $Z_n(X/T) \subset X \times_T \text{Hilb}^n(X/T)$ be the universal subscheme. We consider the diagram

$$
\begin{array}{ccc}
& Z_n(X/T) & \\
{}^{p}\swarrow & & \searrow {}^{q_n} \\
X & & \text{Hilb}^n(X/T) \\
\searrow {}^{\pi} & & \swarrow {}^{\pi_n} \\
& T, &
\end{array}
$$

in which p and q_n are the projections. We get a natural morphism of locally free sheaves

$$r_{n,\mathcal{L}} : \pi_n^*\pi_*(\mathcal{L}) \longrightarrow (q_n)_*p^*(\mathcal{L})$$

on $\text{Hilb}^n(X/T)$ a follows: let

$$f_1 : \pi_n^*\pi_*(\mathcal{L}) \longrightarrow \pi_n^*\pi_*p_*p^*(\mathcal{L}),$$
$$f_2 : \pi_n^*(\pi_n)_*(q_n)_*p^*(\mathcal{L}) \longrightarrow (q_n)_*p^*(\mathcal{L})$$

be the natural morphisms of locally free sheaves on $\text{Hilb}^n(X/T)$. By the commutativity of the diagram we have

$$\pi_n^*\pi_*p_*p^*(\mathcal{L}) = \pi_n^*(\pi_n)_*(q_n)_*p^*(\mathcal{L}),$$

and $r_{n,\mathcal{L}}$ is given by

$$
\begin{array}{ccc}
\pi_n^*\pi_*(\mathcal{L}) & \xrightarrow{\quad r_{n,\mathcal{L}} \quad} & (q_n)_*p^*(\mathcal{L}) \\
\searrow {}^{f_1} & & \nearrow {}^{f_2} \\
\pi_n^*\pi_*p_*p^*(\mathcal{L}) & = & \pi_n^*(\pi_n)_*(q_n)_*p^*(\mathcal{L})
\end{array}
$$

For a fixed $t \in T$ let X_t be the fibre of X over t and put $\mathcal{L}_t := \mathcal{L}|_{X_t}$. For a fixed subscheme $Z \in \text{Hilb}^n(X/T)$ lying in the fibre $\text{Hilb}^n(X/T)_t = \text{Hilb}^n(X_t)$ of $\text{Hilb}^n(X/T)$ over t the map $r_{n,\mathcal{L}}$ between the fibres

$$\pi_n^*\pi_*(\mathcal{L})(Z) = H^0(X_t, \mathcal{L}_t),$$
$$(q_n)_*p^*(\mathcal{L})(Z) = H^0(X_t, \mathcal{L}_t \otimes_{\mathcal{O}_{X_t}} \mathcal{O}_Z)$$

is just given by

$$r_{Z,\mathcal{L}_t} : H^0(X_t, \mathcal{L}_t) \longrightarrow H^0(X_t, \mathcal{L}_t \otimes_{\mathcal{O}_{X_t}} \mathcal{O}_Z).$$

$\mathcal{L}$ is called *n-very ample on X relative to* π, if $r_{m,\mathcal{L}}$ is onto for all $m \leq n+1$ (in other words if for $t \in T$ and all subschemes $Z \subset X_t$ of length $len(Z) \leq n+1$ the map $r_{Z,\mathcal{L}_t}$ is onto).

Remark 4.1.5. Let $\mathcal{L}$ be an $(n-1)$-very ample invertible sheaf on X relative to π. Then $(q_n)_* p^*(\mathcal{L})$ is a locally free quotient of rank n of $\pi_n^* \pi_*(\mathcal{L})$. By the universal property of $Grass(n, \pi_*(\mathcal{L}))$ there is a morphism

$$\phi_{\mathcal{L},n} : \text{Hilb}^n(X/T) \longrightarrow Grass(n, \pi_*(\mathcal{L}))$$

over T such that $\phi_{\mathcal{L},n}^*(Q_{n,\pi_*(\mathcal{L})}) = (q_n)_* p^*(\mathcal{L})$.

As an obvious corollary of theorem 4.1.3 we get :

Remark 4.1.6. Let $\mathcal{L}$ be an n-very ample invertible sheaf on X relative to π. Then $\phi_{\mathcal{L},n}$ is one to one.

The question whether $\phi_{\mathcal{L},n}$ is an embedding we only want to consider in a very simple case.

Definition 4.1.7. Let $X \xrightarrow{\pi} T$ be a locally trivial fibre bundle with fibre X_t and $\mathcal{L}$ an invertible sheaf on X. We call $\mathcal{L}$ *constant over* T, if there is an invertible sheaf $\mathcal{L}_t$ on X_t and an open cover (U_i) of T such that $\pi^{-1}(U_i) \cong U_i \times X_t$ and $\mathcal{L}|_{\pi^{-1}(U_i)} = p_2^*(\mathcal{L}_t)$ with respect to the projection $p_2 : U_i \times X_t \longrightarrow X_t$.

Proposition 4.1.8. *Let* $\mathcal{L}$ *be an* $(n-1)$-*very ample invertible sheaf on* X, *constant over* T. *Then* $\phi_{\mathcal{L},n} : \text{Hilb}^n(X/T) \longrightarrow Grass(n, \pi_*(\mathcal{L}))$ *is an embedding if and only if* $\mathcal{L}$ *is* n-*very ample.*

Proof: As $\mathcal{L}$ is constant over T we have with respect to a suitable local trivialisation $\pi^{-1}(U_i) \cong U_i \times X_t$:

$$\phi_{n,\mathcal{L}}|_{\pi^{-1}(U_i)} = 1_{U_i} \times \phi_{n,\mathcal{L}_t} : U_i \times (X_t)^{[n]} \longrightarrow U_i \times Grass(n, H^0(X_t, \mathcal{L}_t)).$$

The result follows by theorem 4.1.3. $\square$

Now we want to consider the case of the projectivization of a vector bundle. Let E be a vector bundle of rank $d+1$ over a smooth projective variety X. Let $\mathbf{P}(E) \xrightarrow{p} X$ be the bundle of one-dimensional linear subspaces of E and $\mathcal{O}_{\mathbf{P}(E)}(-1) := T_{1,E}$ the tautological subbundle of $p^*(E)$. Let $\check{\mathbf{P}}(E) \xrightarrow{\pi} X$ be the bundle of one-dimensional quotients of E and $Q_{1,E}$ the universal quotient bundle of $\pi^*(E)$. We note that dualizing gives a natural isomorphism $d : \mathbf{P}(E) \longrightarrow \check{\mathbf{P}}(E^*)$ with $d^*(Q_{1,E^*}) = \mathcal{O}_{\mathbf{P}(E)}(1)$. For $Y = \mathbf{P}(E)$ and $Y = \check{\mathbf{P}}(E)$ respectively we again have the projections

$$Z_n(Y/X)$$

$$\swarrow p \qquad\qquad \searrow q_n$$

$$Y \qquad\qquad\qquad\qquad \mathrm{Hilb}^n(Y/X).$$

Proposition 4.1.9.

(1) $Q_{1,E}^{\otimes m}$ *is an m-very ample invertible sheaf on $\check{\mathbf{P}}(E)$ constant over X. For $m \geq n-1$ it gives morphisms*

$$\phi_{n,m} := \phi_{Q_{1,E}^{\otimes m},n} : \mathrm{Hilb}^n(\check{\mathbf{P}}(E)/X) \longrightarrow \mathrm{Grass}(n, \mathrm{Sym}^m(E))$$

over X with $\phi_{n,m}^*(Q_{n,\mathrm{Sym}^m(E)}) = (q_n)_*p^*(Q_{1,E}^{\otimes m})$.

(2) $\phi_n := \phi_{n,n}$ *is an embedding.*

Proof: With respect to a suitable local trivialisation of E over X we have $\pi^{-1}(U_i) = U_i \times \mathbf{P}_d$ and $Q_{1,E}^{\otimes n}|_{\pi^{-1}(U_i)} = p_2^*(\mathcal{O}_{\mathbf{P}_d}(n))$, where $p_2 : U_i \times \mathbf{P}_d \longrightarrow \mathbf{P}_d$ is the projection. (1) follows by $\pi_*(Q_{1,E}^{\otimes m}) = \mathrm{Sym}^m(E)$. (2) follows immediately from 4.1.8 and (1). $\square$

Notation. In future we will write ϕ_n instead of $\phi_{Q_{n,E}^{\otimes n},n}$ and more generally $\phi_{m,n}$ for $\phi_{Q_{n,E}^{\otimes m},n}$, if X and E are understood and $m \geq n-1$.

Now we specialize further to the case that E is a vector bundle of rank 2 on X, i.e. $\mathbf{P}(E)$ is a $\mathbf{P}_1$-bundle over X.

We can express the class $\phi_n^*(c_1(\mathcal{O}_{\mathbf{P}(\mathrm{Sym}^n(E))}(1)))$ in a different way so that its geometric meaning is more visible.

Notation. Let $H_n := (q_n)_*p^*(c_1(\mathcal{O}_{\mathbf{P}(E)}(1))) \in A^1(\mathrm{Hilb}^n(\mathbf{P}(E)/X)$.

Remark 4.1.10. Let $D := \sum a_i D_i$ be a divisor on $\mathbf{P}(E)$ (D_i irreducible, $a_i \in \mathbb{Z}$). Then $(q_n)_*p^*(D) = \sum a_i (q_n)_*p^*(D_i)$, and

$$(q_n)_*p^*(D_i) = \Big\{ Z \in \mathrm{Hilb}^n(\mathbf{P}(E)/X) \,\Big|\, supp(Z) \cap D_i \neq \emptyset \Big\}.$$

Proposition 4.1.11. $\phi_n : \mathrm{Hilb}^n(\mathbf{P}(E)/X) \longrightarrow \mathbf{P}(\mathrm{Sym}^n(E))$ *is an isomorphism such that* $\phi_n^*(c_1(\mathcal{O}_{\mathbf{P}(\mathrm{Sym}^n(E))}(1))) = H_n$.

Proof: As $\mathbf{P}(E)$ is a locally trivial $\mathbf{P}_1$-bundle over X, $\mathrm{Hilb}^n(\mathbf{P}(E)/X)$ has to be a locally trivial $\mathbf{P}_n$-bundle over X. The same is true for $\mathbf{P}(\mathrm{Sym}^n(E))$. So the embedding $\phi_n : \mathrm{Hilb}^n(\mathbf{P}(E)/X) \longrightarrow \mathbf{P}(\mathrm{Sym}^n(E))$ over X must be an isomorphism. Let $x \in X$ and let u, v be a basis of the fibre $E(x)$ of E over x. Then the polynomials of degree n in u, v are in a natural way a basis of the fibre $\mathrm{Sym}^n(E(x)) = \mathrm{Sym}^n(E)(x)$. Let s be a (rational) section of $\mathcal{O}_{\mathbf{P}(E)}(1)$. The application

$$(a_1 u + b_1 v) \cdot \ldots \cdot (a_n u + b_n v) \longmapsto s(a_1 u + b_1 v) \cdot \ldots \cdot s(a_n u + b_n v)$$

gives a (rational) section t of $\mathcal{O}_{\mathbf{P}(\mathrm{Sym}^n(E))}(1)$ with $[div(t)] = H_n$.   □

As the Chern classes of symmetric powers of vector bundles of rank 2 are easy to compute, we know now the Chow ring of $\mathrm{Hilb}^n(\mathbf{P}(E)/X)$. In particular we obtain:

Corollary 4.1.12. *If E is a vector bundle of rank 2 over X with Chern classes* c_1, c_2, *then*

$$A^*(\mathrm{Hilb}^2(\mathbf{P}(E)/X)) = \frac{A^*(X)[H_2]}{(H_2^3 + 3c_1 H_2^2 + (2c_1^2 + 4c_2)H_2 + 4c_1 c_2)}.$$

As a subscheme of length n of a fibre $\mathbf{P}_1$ of $\mathbf{P}(E)$ is just an effective zero cycle of degree n on this fibre, we see that $\mathrm{Hilb}^n(\mathbf{P}(E)/X)$ is the nth symmetric power $\mathrm{Sym}^n(\mathbf{P}(E)/X)$ of $\mathbf{P}(E)$ i.e. the quotient of

$$(\mathbf{P}(E)/X)^n := \mathbf{P}(E) \times_X \mathbf{P}(E) \times_X \ldots \times_X \mathbf{P}(E)$$

by the action of the symmetric group $G(n)$ by permuting the factors. So we have

$$\mathrm{Sym}^n(\mathbf{P}(E)/X) = \mathbf{P}(\mathrm{Sym}^n(E)).$$

Let $Z_n(\mathbf{P}(E)/X) \subset \mathbf{P}(E) \times_X \mathrm{Hilb}^n(\mathbf{P}(E)/X)$ be the universal subscheme. We see from the definitions that $Z_n(\mathbf{P}(E)/X)$ is the reduced subscheme

$$Z_n(\mathbf{P}(E)/X) = \left\{ (x, Z) \in \mathbf{P}(E) \times_X \mathrm{Hilb}^n(\mathbf{P}(E)) \;\middle|\; x \in Z \right\}.$$

We have a natural morphism

$$\psi : \mathbf{P}(E) \times_X \mathrm{Hilb}^{n-1}(\mathbf{P}(E)/X) \longrightarrow \mathrm{Hilb}^n(\mathbf{P}(E)/X).$$

If we identify $\mathrm{Hilb}^n(\mathbf{P}(E)/X)$ with $\mathrm{Sym}^n(\mathbf{P}(E)/X)$, then this morphism is given by $(x, \xi) \longmapsto [x] + \xi$. So we have a morphism

$$p_1 \times \psi : \mathbf{P}(E) \times_X \mathrm{Hilb}^{n-1}(\mathbf{P}(E)/X) \longrightarrow \mathbf{P}(E) \times_X \mathrm{Hilb}^n(\mathbf{P}(E)/X),$$

and we see from the definitions that it is an isomorphism onto its image $Z_n(\mathbf{P}(E)/X)$. If we identify $\mathrm{Hilb}^n(\mathbf{P}(E)/X)$ and $\mathbf{P}(\mathrm{Sym}^n(E))$ then

$$\psi : \mathbf{P}(E) \times_X \mathbf{P}(\mathrm{Sym}^{n-1}(E)) \longrightarrow \mathbf{P}(\mathrm{Sym}^n(E)),$$

is the morphism induced by the natural vector bundle morphism

$$E \otimes \mathrm{Sym}^{n-1}(E) \longrightarrow \mathrm{Sym}^n(E);$$

$$(e \otimes (e_1 \cdot e_2 \cdot \ldots \cdot e_{n-1})) \longmapsto (e \cdot e_1 \cdot e_2 \cdot \ldots \cdot e_{n-1})$$

So we get:

Lemma 4.1.13.

$$1_{\mathbf{P}(E)} \times \phi_n : \mathbf{P}(E) \times_X \mathrm{Hilb}^n(\mathbf{P}(E)/X) \longrightarrow \mathbf{P}(E) \times_X \mathbf{P}(\mathrm{Sym}^n(E))$$

induces an isomorphism

$$\widetilde{\psi} : Z_n(\mathbf{P}(E)/X) \longrightarrow \mathbf{P}(E) \times_X \mathbf{P}(\mathrm{Sym}^{n-1}(E)).$$

We see that with respect to the projections p_1, p_2 of $\mathbf{P}(E) \times_X \mathbf{P}(\mathrm{Sym}^{n-1}(E))$ to $\mathbf{P}(E)$ and $\mathbf{P}(\mathrm{Sym}^{n-1}(E))$ we have

$$\widetilde{\psi}^*(\mathcal{O}_{\mathbf{P}(\mathrm{Sym}^n(E))}(1)) = p_1^*(\mathcal{O}_{\mathbf{P}(E)}(1)) \otimes p_2^*(\mathcal{O}_{\mathbf{P}(\mathrm{Sym}^{n-1}(E))}(1)).$$

Now let E be a vector bundle of arbitrary rank $d+1$ over X.

Definition 4.1.14. Let $Al^n(\mathbf{P}(E))$ be the reduced subvariety of $\mathrm{Hilb}^n(\mathbf{P}(E)/X)$, given by

$$Al^n(\mathbf{P}(E))) = \left\{ Z \in \mathrm{Hilb}^n(\mathbf{P}(E)/X) \,\middle|\, \begin{array}{c} Z \text{ is a subscheme of a line} \\ \text{in a fibre } \mathbf{P}_d \end{array} \right\}.$$

Let $Z_n^{al}(\mathbf{P}(E))$ be the universal subscheme over $Al^n(\mathbf{P}(E))$ and let

$$Z_n^{al}(\mathbf{P}(E))$$

$$\swarrow {\scriptstyle \tilde{p}} \qquad\qquad\qquad \searrow {\scriptstyle q_n}$$

$$\mathbf{P}(E) \qquad\qquad\qquad\qquad Al^n(\mathbf{P}(E))$$

be the projections. In particular let $Al^n(\mathbf{P}_d) \subset \mathbf{P}_d^{[n]}$ be the subvariety given by

$$Al^n(\mathbf{P}_d) = \left\{ Z \in \mathbf{P}_d^{[n]} \;\middle|\; Z \text{ is a subscheme of a line in } \mathbf{P}_d \right\}$$

and $Z_n^{al}(\mathbf{P}_d)$ the universal subscheme over $Al^n(\mathbf{P}_d)$.

Let $H, L_{n-1}, H_n \in A^1(Z_n^{al}(\mathbf{P}(E)))$ be the classes defined by

$$H := \bar{p}^*(c_1(\mathcal{O}_{\mathbf{P}(E)}(1))),$$
$$H_n := \bar{q}_n^*(\bar{q}_n)_*(H),$$
$$L_{n-1} := H_n - H.$$

We will also denote by H_n the class $(\bar{q}_n)_*(H) \in A^1(Al^n(\mathbf{P}(E)))$.

Let $G := Grass(d-1, E)$, which we view as the variety of lines in the fibres of $\mathbf{P}(E)$. Let $T := T_{2,E}$ be the tautological bundle of rank 2 over G. We can associate to each subscheme $Z \in Al^n(\mathbf{P}(E))$ the line on which it lies. It is easy to see that this defines a morphism

$$axe : Al^n(\mathbf{P}(E)) \longrightarrow G.$$

Let $F \subset \mathbf{P}(E) \times_X G$ be the incidence variety

$$F := \left\{ (x, l) \in \mathbf{P}(E) \times_X G \;\middle|\; x \in l \right\}$$

with the projections

$$F$$
$$\swarrow \bar{p}_1 \qquad \searrow \bar{p}_2$$
$$\mathbf{P}(E) \qquad\qquad G.$$

Then we can identify $F \xrightarrow{\bar{p}_2} G$ with $\mathbf{P}(T) \longrightarrow G$, and with this identification we have $\mathcal{O}_{\mathbf{P}(T)}(1) = \bar{p}_1^*(\mathcal{O}_{\mathbf{P}(E)}(1))$. Obviously the relative Hilbert scheme

$$\mathrm{Hilb}^n(F/G) \subset \mathrm{Hilb}^n(\mathbf{P}(E)/X) \times_X G$$

is the closed reduced subscheme

$$\mathrm{Hilb}^n(F/G) = \left\{ (Z, l) \in \mathrm{Hilb}^n(\mathbf{P}(E)/X) \times_X G \;\middle|\; Z \subset l \right\},$$

where we have now identified the points of G with the lines l in the fibres of E. We see that the projection $\tilde{p}_1 : \mathrm{Hilb}^n(F/G) \longrightarrow \mathrm{Hilb}^n(\mathbf{P}(E)/X)$ defines an isomorphism of $\mathrm{Hilb}^n(F/G)$ onto its image $Al^n(\mathbf{P}(E))) \subset \mathrm{Hilb}^n(\mathbf{P}(E)/X)$. (As a morphism to $Al^n(\mathbf{P}(E)))$ it is obviously a bijection, and both $\mathrm{Hilb}^n(F/G)$ and $Al^n(\mathbf{P}(E))$ are smooth). Let

$$Z_n(F/G) \subset \mathbf{P}(E) \times \mathrm{Hilb}^n(\mathbf{P}(E)/X) \times G$$

be the universal subscheme. We see that the projection $\widetilde{p}_{1,2} : Z_n(F/G) \longrightarrow$ $Z_n(\mathbf{P}(E)/X)$ gives an isomorphism of $Z_n(F/G)$ onto $Z_n^{al}(\mathbf{P}(E))$. So we get by lemma 4.1.13:

Lemma 4.1.15.

(1) $$\widetilde{\phi}_n = \phi_n {\circ} \widetilde{p}_1^{-1} : Al^n(\mathbf{P}(E)) \longrightarrow \mathbf{P}(\mathrm{Sym}^n(T))$$

is an isomorphism over G, such that $\widetilde{\phi}_n^*(c_1(\mathcal{O}_{\mathbf{P}(\mathrm{Sym}^n(T))}(1))) = H_n$.

(2) $$\widetilde{\phi}_{1,n} := \widetilde{\psi} {\circ} \widetilde{p}_{1,2}^{-1} : Z_n^{al}(\mathbf{P}(E)) \longrightarrow \mathbf{P}(T) \times_G \mathbf{P}(\mathrm{Sym}^{n-1}(T))$$

is an isomorphism satisfying

$$\widetilde{\phi}_{1,n}^*(p_1^*(c_1(\mathcal{O}_{\mathbf{P}(T)}(1)))) = H$$
$$\widetilde{\phi}_{1,n}^*(p_2^*(c_1(\mathcal{O}_{\mathbf{P}(\mathrm{Sym}^{n-1}(T))}(1)))) = L_{n-1}.$$

So by proposition 4.1.11 we now know the Chow ring of $Al^n(\mathbf{P}_d)$. We keep in mind that by remark 4.1.10 we can write the class $H_n \in A^1(Al^n(\mathbf{P}_d))$ as

$$H_n = \left[\Big\{ Z \in Al^n(\mathbf{P}_d) \mid supp(Z) \text{ intersects a fixed hyperplane} \Big\}\right].$$

So we get:

Example 4.1.16.

$$A^*(Al^n(\mathbf{P}_d)) = \frac{A^*(Grass(d-1,d+1))[H_n]}{\left(\displaystyle\sum_{i=0}^{n+1} c_i(\mathrm{Sym}^n(T_{2,d+1}))H_n^{n+1-i}\right)}.$$

In particular we have with $P := c_1(Q_{1,3})$:

$$A^*(Al^n(\mathbf{P}_2)) = \frac{\mathbb{Z}[P, H_n]}{\left(P^3,\ H_n^{n+1} - \binom{n+1}{2}H_n^n P + w(n)H_n^{n-1}P^2\right)}.$$

Here

$$w(n) = \begin{cases} \frac{n(2n+1)(n+1)}{6} + \frac{(3n^2-2n)(n^2-1)}{24}, & n \text{ odd;} \\ \frac{n(2n+1)(n+1)}{6} + \frac{(n-2)(n-1)n}{24} + \frac{n^3(n-1)}{8}, & n \text{ even.} \end{cases}$$

4.2. Computation of the Chow ring of $\widetilde{\mathrm{Hilb}}^3(\mathbf{P}_2)$

Now we want to use the results of the preceeding section to compute the Chow ring of the variety $\widetilde{\mathrm{Hilb}}^3(\mathbf{P}_2)$ of triangles in $\mathbf{P}_2$ with a marked side. Remember that $\widetilde{\mathrm{Hilb}}^3(\mathbf{P}_2) \subset \mathbf{P}_2^{[2]} \times \mathbf{P}_2^{[3]}$ is defined as the subvariety

$$\widetilde{\mathrm{Hilb}}^3(\mathbf{P}_2) := \left\{ (Z_2, Z_3) \in \mathbf{P}_2^{[2]} \times \mathbf{P}_2^{[3]} \mid Z_2 \subset Z_3 \right\}.$$

$\widetilde{\mathrm{Hilb}}^3(\mathbf{P}_2)$ was defined in [Elencwajg-Le Barz (2),(3)] to compute the Chow ring of $\mathbf{P}_2^{[3]}$. The result is however quite complicated. In this section we shall see that the Chow ring of $\widetilde{\mathrm{Hilb}}^3(\mathbf{P}_2)$ is relatively simple, so it might be more useful for some enumerative applications. If the ground field is $\mathbf{C}$, then the Chow ring of $\widetilde{\mathrm{Hilb}}^3(\mathbf{P}_2)$ coincides with the cohomology ring (Proposition 2.5.19). Let $res : \widetilde{\mathrm{Hilb}}^3(\mathbf{P}_2) \longrightarrow \mathbf{P}_2$ be the residual morphism (see lemma 2.5.3) and

$$\widetilde{\mathrm{Hilb}}^3(\mathbf{P}_2)$$

$$\swarrow \bar{p}_1 \qquad\qquad \searrow \bar{p}_2$$

$$\mathbf{P}_2^{[2]} \qquad\qquad\qquad\qquad \mathbf{P}_2^{[3]}$$

the projections. By proposition 2.5.19 we get

$$A^1(\widetilde{\mathrm{Hilb}}^3(\mathbf{P}_2)) = A^5(\widetilde{\mathrm{Hilb}}^3(\mathbf{P}_2)) = \mathbb{Z}^4$$
$$A^2(\widetilde{\mathrm{Hilb}}^3(\mathbf{P}_2)) = A^4(\widetilde{\mathrm{Hilb}}^3(\mathbf{P}_2)) = \mathbb{Z}^9$$
$$A^3(\widetilde{\mathrm{Hilb}}^3(\mathbf{P}_2)) = \mathbb{Z}^{11}.$$

Now we define some elements of $A^1(\widetilde{\mathrm{Hilb}}^3(\mathbf{P}_2))$, which will generate the Chow ring of $\widetilde{\mathrm{Hilb}}^3(\mathbf{P}_2)$.

Definition 4.2.1. Let $Z_2(\mathbf{P}_2) \subset \mathbf{P}_2 \times \mathbf{P}_2^{[2]}$ be the universal subscheme and let

$$Z_2(\mathbf{P}_2)$$

$$\swarrow p \qquad\qquad \searrow q_2$$

$$\mathbf{P}_2 \qquad\qquad\qquad\qquad \mathbf{P}_2^{[2]}$$

be the projections. Let $H := res^*(c_1(\mathcal{O}_{\mathbf{P}_2}(1)))$ and let

$$axe : \mathbf{P}_2^{[2]} = Al^2(\mathbf{P}_2) \longrightarrow \check{\mathbf{P}}_2$$

be the axial morphism of 4.1.14. We put

$$P := \bar{p}_1^* axe^*(c_1(Q_{1,3})),$$
$$H_2 := \bar{p}_1^*(q_2)_* p^*(c_1(\mathcal{O}_{\mathbf{P}_2}(1))).$$

Let $\tilde{A} \subset \widetilde{\mathrm{Hilb}}^3(\mathbf{P}_2)$ be the subvariety

$$\tilde{A} := \left\{ (Z_2, Z_3) \in \widetilde{\mathrm{Hilb}}^3(\mathbf{P}_2) \,\middle|\, Z_3 \text{ is a subscheme of a line} \right\}$$

and $A := [\tilde{A}] \in A^1(\widetilde{\mathrm{Hilb}}^3(\mathbf{P}_2))$. Let $\tilde{P}_2 \subset \widetilde{\mathrm{Hilb}}^3(\mathbf{P}_2)$ be the closed subvariety

$$\tilde{P}_2 := \left\{ (Z_2, Z_3) \in \widetilde{\mathrm{Hilb}}^3(\mathbf{P}_2) \,\middle|\, \begin{array}{c} \text{the line through one of the subschemes } Z_1 \subset Z_3 \\ \text{of length 2 containing } res(Z_2, Z_3) \\ \text{passes through a fixed point} \end{array} \right\}$$

and $P_2 := [\tilde{P}_2]$.

Remark 4.2.2. Geometrically H_2, H, P can also be described as

$$H_2 = \left[\left\{ (Z_2, Z_3) \in \widetilde{\mathrm{Hilb}}^3(\mathbf{P}_2) \,\middle|\, \begin{array}{c} \text{a point of } Z_2 \\ \text{lies on a fixed line} \end{array} \right\} \right],$$

$$H = \left[\left\{ (Z_2, Z_3) \in \widetilde{\mathrm{Hilb}}^3(\mathbf{P}_2) \,\middle|\, res(Z_2, Z_3) \text{ lies on a fixed line} \right\} \right],$$

$$P = \left[\left\{ (Z_2, Z_3) \in \widetilde{\mathrm{Hilb}}^3(\mathbf{P}_2) \,\middle|\, \begin{array}{c} \text{the line through } Z_2 \\ \text{passes through a fixed point} \end{array} \right\} \right].$$

Theorem 4.2.3.
$$A^*(\widetilde{\mathrm{Hilb}}^3(\mathbf{P}_2)) = \frac{\mathbb{Z}[H, H_2, P, A]}{(I_1, I_2, I_3, I_4, I_5, I_6)}$$

with

$$I_1 := H^3,$$
$$I_2 := P^3,$$
$$I_3 := H_2^3 - 3H_2^2 P + 6H_2 P^2,$$
$$I_4 := A(H^2 - HP + P^2),$$
$$I_5 := A(A - 3P + H + H_2),$$
$$I_6 := AH_2^2 - (H_2^2 P - H_2 P^2 + HH_2^2 - 3HH_2 P + 2HP^2 - 2H^2 H_2 + 2H^2 P$$
$$+ AH_2 P + 2AHH_2 - AHP).$$

Proof: By example 4.1.16 the subring of $A^*(\widetilde{\mathrm{Hilb}}^3(\mathbf{P}_2))$ generated by H, P, H_2 is

$$(res \times \bar{p}_1)^*(A^*(\mathbf{P}_2 \times \mathbf{P}_2^{[2]})) = \frac{\mathbb{Z}[H, P, H_2]}{(I_1, I_2, I_3)}.$$

As the morphism $res \times \bar{p}_1 : \widetilde{\mathrm{Hilb}}^3(\mathbf{P}_2) \longrightarrow \mathbf{P}_2 \times \mathbf{P}_2^{[2]}$ is birational, the orientation cycle of $\widetilde{\mathrm{Hilb}}^3(\mathbf{P}_2)$ is the class $[*] := H^2 H_2^2 P^2$. The restriction of $res \times \bar{p}_2$ to the subvariety $\widetilde{A}$ gives an isomorphism $\psi : \widetilde{A} \longrightarrow Z_3^{al}(\mathbf{P}_2) \subset Z_3(\mathbf{P}_2)$. By lemma 4.1.15 we have

$$Z_3^{al}(\mathbf{P}_2) = \mathbf{P}(T_{2,3}) \times_{\check{\mathbf{P}}_2} \text{Hilb}^2(\mathbf{P}(T_{2,3})/\check{\mathbf{P}}_2),$$

where $T_{2,3}$ is the tautological bundle over $\check{\mathbf{P}}_2 = Grass(1,3)$. So we get

$$A^*(\widetilde{A}) = \frac{\mathbb{Z}[H, H_2, P]}{(P^3, H^2 - HP + P^2, H_2^3 - 3H_2^2 P + 6H_2 P^2)},$$

and the orientation cycle of $\widetilde{A}$ is $P^2 H H_2^2$. So relation $I_4 = 0$ holds in $A^*(\widetilde{\text{Hilb}}^3(\mathbf{P}_2))$, and for the orientation cycle we get $[*] = A H H_2^2 P^2$. To show $I_5 = 0$ we use the class $P_2 \in A^1(\widetilde{\text{Hilb}}^3(\mathbf{P}_2))$.

Lemma 4.2.4. $P + P_2 = A + H + H_2$.

Proof: Let

$$\overline{A} := \left[\left\{ Z \in \mathbf{P}_2^{[3]} \;\middle|\; Z \text{ lies on a line} \right\} \right],$$

$$\overline{H} := \left[\left\{ Z \in \mathbf{P}_2^{[3]} \;\middle|\; Z \text{ intersects a fixed line} \right\} \right],$$

$$\overline{P} := \left[\left\{ Z \in \mathbf{P}_2^{[3]} \;\middle|\; \begin{array}{l} \text{a subscheme } Z_2 \text{ of length 2 of } Z \\ \text{lies on a line passing through a fixed point} \end{array} \right\} \right].$$

So we have by definition $\overline{H} = (\bar{p}_2)_*(H)$, $\overline{P} = (\bar{p}_2)_*(P)$, $A = \bar{p}_2^*(\overline{A})$, and we see that the relations

$$\bar{p}_2^*(\overline{H}) = H + H_2,$$
$$\bar{p}_2^*(\overline{P}) = P + P_2,$$
$$(\bar{p}_2)_*(A) = 3\overline{A}$$

hold, as the projection $\bar{p}_2 : \widetilde{\text{Hilb}}^3(\mathbf{P}_2) \longrightarrow \mathbf{P}_2^{[3]}$ is generically finite of degree 3. In [Elencwajg-Le Barz (3)] it has been shown that the relation $\overline{P} = \overline{A} + \overline{H}$ holds in $A^1(\mathbf{P}_2^{[3]})$. We briefly repeat the elementary argument: we put

$$\phi_1 := (\bar{p}_2)_*(H H_2^2 P^2),$$
$$\phi_2 := (\bar{p}_2)_*(H^2 H_2 P^2) \in A^5(\mathbf{P}_2^{[3]}).$$

These classes can be geometrically described as follows:

$$\phi_1 = \left[\left\{ Z \in \mathbf{P}_2^{[3]} \;\middle|\; \begin{array}{l} Z \text{ consists of two distinct fixed points} \\ x_1, x_2 \text{ and another point } x_3 \text{ moving on a} \\ \text{fixed line containing neither } x_1 \text{ nor } x_2. \end{array} \right\} \right],$$

$$\phi_2 = \left[\left\{ Z \in \mathbf{P}_2^{[3]} \;\middle|\; \begin{array}{l} Z \text{ consists of a fixed point } x \text{ and} \\ \text{a subscheme } Z_2 \text{ of length 2 on a} \\ \text{fixed line } l \text{ not containing } x; \\ Z_2 \text{ contains a fixed point } x_2 \in l. \end{array} \right\} \right].$$

Using this description we can easily compute the intersection table:

	$\overline{H}$	$\overline{A}$	$\overline{P}$
ϕ_1	1	1	2
ϕ_2	1		1

As the group $A^1(\mathbf{P}_2^{[3]}) = A^5(\mathbf{P}_2^{[3]})$ is free of rank 2, we see that $\overline{H}, \overline{A}$ and ϕ_1, ϕ_2 form bases of $A^1(\mathbf{P}_2^{[3]})$ and $A^5(\mathbf{P}_2^{[3]})$ respectively and the relation $\overline{P} = \overline{A} + \overline{H}$ holds. The result follows. $\quad\square$

Lemma 4.2.5. $AP_2 = 2AP$.

Proof: We have to show the relation $P_2|_{\widetilde{A}} = 2P|_{\widetilde{A}}$. We have

$$\widetilde{A} \cong \mathbf{P}(T_{2,3}) \times_{\check{\mathbf{P}}_2} \mathbf{P}(\mathrm{Sym}^2(T_{2,3})).$$

Let

$$\pi_1 : \mathbf{P}(T_{2,3}) \longrightarrow \check{\mathbf{P}}_2$$
$$\pi_2 : \mathbf{P}(\mathrm{Sym}^2(T_{2,3})) \longrightarrow \check{\mathbf{P}}_2$$
$$p_1 : \mathbf{P}(T_{2,3}) \times_{\check{\mathbf{P}}_2} \mathbf{P}(\mathrm{Sym}^2(T_{2,3})) \longrightarrow \mathbf{P}(T_{2,3})$$
$$p_2 : \mathbf{P}(T_{2,3}) \times_{\check{\mathbf{P}}_2} \mathbf{P}(\mathrm{Sym}^2(T_{2,3})) \longrightarrow \mathbf{P}(\mathrm{Sym}^2(T_{2,3}))$$

be the projections. Then we have $P = p_2^*(\pi_2^*(c_1(Q_{1,3})))$. Let

$$\widehat{A} := \mathbf{P}(T_{2,3}) \times_{\check{\mathbf{P}}_2} \mathbf{P}(T_{2,3}) \times_{\check{\mathbf{P}}_2} \mathbf{P}(T_{2,3}),$$

where $\bar{p}_1, \bar{p}_2, \bar{p}_3 : \widehat{A} \longrightarrow \mathbf{P}(T_{2,3})$ are the projections. We consider the natural morphism $\phi : \widehat{A} \longrightarrow \mathbf{P}(T_{2,3}) \times_{\check{\mathbf{P}}_2} \mathbf{P}(\mathrm{Sym}^2(T_{2,3}))$. Let

$$\widetilde{\pi} : \mathbf{P}(T_{2,3}) \times_{\check{\mathbf{P}}_2} \mathbf{P}(T_{2,3}) \longrightarrow \check{\mathbf{P}}_2$$
$$\widetilde{\widetilde{\pi}} : \mathbf{P}(T_{2,3}) \times_{\check{\mathbf{P}}_2} \mathbf{P}(T_{2,3}) \times_{\check{\mathbf{P}}_2} \mathbf{P}(T_{2,3}) \longrightarrow \check{\mathbf{P}}_2$$

be the projections. Then we see

$$\phi^*(P) = (\bar{p}_2 \times \bar{p}_3)^*(\widetilde{\pi}^*(c_1(Q_{1,3})))$$
$$= \widetilde{\widetilde{\pi}}^*(c_1(Q_{1,3})),$$
$$\phi^*(P_2) = (\bar{p}_1 \times \bar{p}_2)^*(\widetilde{\pi}^*(c_1(Q_{1,3}))) + (\bar{p}_1 \times \bar{p}_3)^*(\widetilde{\pi}^*(c_1(Q_{1,3})))$$
$$= 2\widetilde{\widetilde{\pi}}^*(c_1(Q_{1,3})),$$
$$\phi_*(\widetilde{\widetilde{\pi}}^*(c_1(Q_{1,3}))) = 2P|_{\widetilde{A}}.$$

The lemma follows. $\quad\square$

From lemmas 4.2.4 and 4.2.5 we get the relation $I_5 = 0$:

$$A^2 = A(P + P_2 - H - H_2)$$
$$= 3AP - AH - AH_2.$$

The information we have obtained until now is already enough to determine the ring structure of $A^*(\widetilde{\mathrm{Hilb}}^3(\mathbf{P}_2))$.

We use relations $I_1, \ldots, I_5$ to compute the intersection tables. We also use that the orientation class is $[*] = AP^2HH_2^2 = H^2H_2^2P^2$. We get the following tables:

$A^1 \times A^5$

	H_2	P	H	A
$HH_2^2P^2$			1	1
$H^2H_2^2P$	3	1		1
$H^2H_2P^2$	1			
AHH_2P^2	1			−1

$A^2 \times A^4$

	H_2^2	H_2P	P^2	HH_2	HP	H^2	AH_2	AP	AH
$H_2^2P^2$						1			1
HH_2^2P				3	1		3	1	1
HH_2P^2				1			1		
$H^2H_2^2$	3	3	1				3	1	
H^2H_2P	3	1					1		
H^2P^2	1								
AH_2P^2				1			−1		−1
AHH_2P	3	1		1			−1	−1	−1
AHP^2	1						−1		

$A^3 \times A^3$

	H_2^2P	H_2P^2	HH_2^2	HH_2P	HP^2	H^2H_2	H^2P	AH_2P	AP^2	AHH_2	AHP
H_2^2P						3	1			3	1
H_2P^2						1				1	
HH_2^2			3	3	1			3	1	3	1
HH_2P			3	1				1		1	
HP^2			1								
H^2H_2	3	1						1			
H^2P	1										
AH_2P			3	1		1		-1		-1	-1
AP^2			1					-1			
AHH_2	3	1	3	1				-1	-1		-1
AHP	1		1					-1		-1	
AH_2^2			3	3	1	3	1	-3	-1	3	-1

We see that the intersection matrices are all invertible over $\mathbb{Z}$. By solving the system of equations given by the last intersection matrix we get $I_6 = 0$.

End of the proof of theorem 4.2.3: As we have found a $\mathbb{Z}$-basis of $A^*(\widetilde{\mathrm{Hilb}}^3(\mathbf{P}_2))$ consisting of monomials in H_2, P, H, A the ring $A^*(\widetilde{\mathrm{Hilb}}^3(\mathbf{P}_2))$ is generated by H_2, P, H, A. We also have seen that the relations $I_1 = 0, \ldots, I_6 = 0$ hold. We have to show that these generate all the relations. For this it is enough to show that every monomial in H_2, P, H, A can be expressed in terms of the elements of the basis by making use of $I_1, \ldots, I_6$. Let M be such a monomial. By $I_1, \ldots, I_6$ it can be expressed as a linear combination of monomials $A^a H^h P^p H_2^{h_2}$ satisfying

$$h \leq 2 \qquad (I_1),$$
$$p \leq 2 \qquad (I_2),$$
$$h_2 \leq 2 \qquad (I_3),$$
$$a \leq 1 \qquad (I_5),$$
$$h + a \leq 2 \qquad (I_4),$$
$$a + h_2 \leq 2 \qquad (I_6).$$

We see that these conditions are only satisfied by the elements of the basis occuring in the above intersection matrices. $\square$

4.3. The Chow ring of $\widetilde{\mathrm{Hilb}}^3(\mathbf{P}(E)/X)$

Now we want to generalize the result of the last section. Let X be a smooth variety and E a vector bundle of rank 3 on X.

Definition 4.3.1. Let $\widetilde{\mathrm{Hilb}}^3(\mathbf{P}(E)/X) \subset \mathrm{Hilb}^2(\mathbf{P}(E)/X) \times_X \mathrm{Hilb}^3(\mathbf{P}(E)/X)$ be the subvariety defined by

$$\widetilde{\mathrm{Hilb}}^3(\mathbf{P}(E)/X) := \Big\{ (Z_1, Z) \in \mathrm{Hilb}^2(\mathbf{P}(E)/X) \times_X \mathrm{Hilb}^3(\mathbf{P}(E)/X) \,\Big|\, Z_1 \subset Z \Big\}.$$

Let

$$V(\mathbf{P}(E)) := \mathbf{P}(E) \times_X \mathbf{P}(E) \times_X \mathbf{P}(E) \times_X \mathrm{Hilb}^2(\mathbf{P}(E)/X) \times_X \mathrm{Hilb}^2(\mathbf{P}(E)/X)$$
$$\times_X \mathrm{Hilb}^2(\mathbf{P}(E)/X) \times_X \mathrm{Hilb}^3(\mathbf{P}(E)/X)$$

and $\widehat{H}^3(\mathbf{P}(E)/X) \subset V(\mathbf{P}(E))$ be the subvariety defined by

$$\widehat{H}^3(\mathbf{P}(E)/X) := \left\{ \begin{array}{c} (x_1, x_2, x_3, Z_1, Z_2, Z_3, Z) \\ \in V(\mathbf{P}(E)) \end{array} \,\middle|\, \begin{array}{c} x_i, x_j \subset Z_k; \ Z_i \subset Z; \\ x_k = res(x_i, Z_j) = res(Z_k, Z) \\ \text{for all permutations} \\ (i, j, k) \text{ of } (1,2,3) \end{array} \right\}.$$

As $\mathrm{Hilb}^n(\mathbf{P}(E)/X)$ is a locally trivial fibre bundle over X with fibre $\mathbf{P}_2^{[n]}$, we see easily:

Remark 4.3.2.

(1) $\widetilde{\mathrm{Hilb}}^3(\mathbf{P}(E)/X)$ is a locally trivial fibre bundle over X with fibre $\widetilde{\mathrm{Hilb}}^3(\mathbf{P}_2)$.

(2) $\widehat{H}^3(\mathbf{P}(E)/X)$ is a locally trivial fibre bundle over X with fibre $\widehat{H}^3(\mathbf{P}_2)$.

$\widetilde{\mathrm{Hilb}}^3(\mathbf{P}(E)/X)$ parametrizes the triangles with a marked side and $\widehat{H}^3(\mathbf{P}(E)/X)$ the complete triangles in the fibres $F \cong \mathbf{P}_2$ of $\mathbf{P}(E)$ over X. We want to use results from [Collino-Fulton (1)] on the Chow ring of $\widehat{H}^3(\mathbf{P}(E)/X)$, to compute $A^*(\widetilde{\mathrm{Hilb}}^3(\mathbf{P}(E)/X))$. In [Collino-Fulton (1)] another definition of $\widehat{H}^3(\mathbf{P}(E)/X)$ is used, which we will denote by $W(\mathbf{P}(E)/X)$. First we give the definition of $W(\mathbf{P}(E)/X)$.

Definition 4.3.3. Let

$$U(\mathbf{P}(E)) := \mathbf{P}(E) \times_X \mathbf{P}(E) \times_X \mathbf{P}(E) \times_X \check{\mathbf{P}}(E) \times_X \check{\mathbf{P}}(E) \times_X$$
$$\check{\mathbf{P}}(E) \times_X Grass(3, \mathrm{Sym}^2(E))$$

and let $s : U(\mathbf{P}(E)) \longrightarrow X$ be the projection. Let $x \in X$.

$$y = (x_1, x_2, x_3, \xi_1, \xi_2, \xi_3, \Gamma) \in s^{-1}(x)$$

is called a honest triangle if x_1, x_2, x_3 are three distinct points of a fibre $\mathbf{P}(E(x))$ and ξ_k is the line connecting x_i, x_j (for all permutations (i, j, k) of $(1, 2, 3)$) and Γ is the linear system of conics passing through x_1, x_2, x_3, viewed as an element of the fibre $Grass(3, \mathrm{Sym}^2(E(x)))$. Let $W_0(\mathbf{P}(E)) \subset U(\mathbf{P}(E))$ be the set of honest triangles and $W(\mathbf{P}(E))$ the closure of $W_0(\mathbf{P}(E))$ in $U(\mathbf{P}(E))$.

Now we want to construct an embedding of $\widehat{H}^3(\mathbf{P}(E)/X)$ into a product of bundles of Grassmannians. By the results of section 4.1 we get that the morphism $\phi_1^3 \times \phi_2^3 \times \phi_3|_{\widehat{H}^3(\mathbf{P}(E)/X)}$ is a closed embedding of $\widehat{H}^3(\mathbf{P}(E)/X)$ into

$$\mathbf{P}(E) \times_X \mathbf{P}(E) \times_X \mathbf{P}(E) \times_X Grass(4, \mathrm{Sym}^2(E)) \times_X Grass(4, \mathrm{Sym}^2(E)) \times_X$$
$$Grass(4, \mathrm{Sym}^2(E)) \times_X Gr(7, \mathrm{Sym}^3(E)).$$

On the other hand in [Le Barz (10)] $\widehat{H}^3(\mathbf{P}_2)$ was shown to be a closed subscheme of $\mathbf{P}_2^3 \times \check{\mathbf{P}}_2^3 \times Grass(3, 6)$, and we can see from the proof that the embedding $\widehat{H}^3(\mathbf{P}_2) \longrightarrow \mathbf{P}_2^3 \times \check{\mathbf{P}}_2^3 \times Grass(3, 6)$ is given by the morphism

$$\Phi := \phi^3_{\mathcal{O}_{\mathbf{P}_2}(1),1} \times \phi^3_{\mathcal{O}_{\mathbf{P}_2}(1),2} \times \phi_{\mathcal{O}_{\mathbf{P}_2}(2),3}|_{\widehat{H}^3(\mathbf{P}_2)}.$$

We have the morphisms

$$1_{\mathbf{P}(E)} = \phi_{\mathcal{O}_{\mathbf{P}(E)}(1),1} : \mathbf{P}(E) \longrightarrow \mathbf{P}(E),$$
$$axe := \phi_{\mathcal{O}_{\mathbf{P}(E)}(1),2} : \mathrm{Hilb}^2(\mathbf{P}(E)/X) \longrightarrow Grass(1, E),$$
$$\phi_{2,3} := \phi_{\mathcal{O}_{\mathbf{P}(E)}(2),3} : \mathrm{Hilb}^3(\mathbf{P}(E)/X) \longrightarrow Grass(3, \mathrm{Sym}^2(E)).$$

Proposition 4.3.4.

$$\widehat{\Phi} := 1^3_{\mathbf{P}(E)} \times axe^3 \times \phi_{2,3} : \widehat{H}^3(\mathbf{P}(E)/X) \longrightarrow U(\mathbf{P}(E))$$

is a closed embedding with image $W(\mathbf{P}(E))$.

Proof: Let $U \subset X$ be an open subset over which E is trivial. Then with respect to suitable local trivialisations over U the restriction of $\widehat{\Phi}$ is the closed embedding

$$1_U \times \Phi : U \times \widehat{H}^3(\mathbf{P}_2) \longrightarrow U \times \mathbf{P}_2^3 \times \check{\mathbf{P}}_2^3 \times Grass(3, 6).$$

So $\widehat{\Phi}$ is a closed embedding. We can see immediately that the image of the open subvariety

$$\widehat{H}^3_{(1,1,1)}(\mathbf{P}(E)) := \left\{ \begin{matrix} (x_1, x_2, x_3, Z_1, Z_2, Z_3, Z) \\ \in \widehat{H}^3(\mathbf{P}(E)/X) \end{matrix} \;\middle|\; \text{the } x_i \text{ are distinct} \right\}$$

is the variety $W_0(\mathbf{P}(E)) \subset U(\mathbf{P}(E))$ of honest triangles in $\mathbf{P}(E)$. As $\widehat{H}^3_{(1,1,1)}(X)$ is open and dense in $\widehat{H}^3(X)$ and $W(\mathbf{P}(E))$ is defined as the closure of $W_0(\mathbf{P}(E))$ in $U(\mathbf{P}(E))$, the result follows. □

In [Collino-Fulton (1)] the Chow ring of $W(\mathbf{P}(E)/X)$ is computed as an algebra over $A^*(X)$. There the following classes are important:

Definition 4.3.5. Let

$$\bar{p}_1, \bar{p}_2, \bar{p}_3 : W(\mathbf{P}(E)) \longrightarrow \mathbf{P}(E),$$
$$\bar{q}_1, \bar{q}_2, \bar{q}_3 : W(\mathbf{P}(E)) \longrightarrow \check{\mathbf{P}}(E),$$
$$\bar{q} : W(\mathbf{P}(E)) \longrightarrow Grass(3, \operatorname{Sym}^2(E))$$

be the projections. We put

$$a := \bar{p}_1^*(c_1(\mathcal{O}_{\mathbf{P}(E)}(1))), \quad b := \bar{p}_2^*(c_1(\mathcal{O}_{\mathbf{P}(E)}(1))), \quad c := \bar{p}_3^*(c_1(\mathcal{O}_{\mathbf{P}(E)}(1))),$$
$$\alpha := \bar{q}_1^*(c_1(T_{2,E}^*)), \quad \beta := \bar{q}_2^*(c_1(T_{2,E}^*)), \quad \gamma := \bar{q}_3^*(c_1(T_{2,E}^*)).$$

Then $a, b, c, \alpha, \beta, \gamma \in A^1(W(\mathbf{P}(E)))$. Let $\pi : \mathbf{P}(E) \longrightarrow X$, $\bar{\pi} : W(\mathbf{P}(E)) \longrightarrow X$ be the projections. We write:

$$\mu_1 := \bar{\pi}^*(c_1(E^*)) = -\bar{\pi}^*(c_1(E)),$$
$$\mu_2 := \bar{\pi}^*(c_2(E^*)) = \bar{\pi}^*(c_2(E)),$$
$$\mu_3 := \bar{\pi}^*(c_3(E^*)) = -\bar{\pi}^*(c_3(E)).$$

Let $\epsilon \in A^1(W(\mathbf{P}(E)))$ be the class of the subvariety

$$\widetilde{\epsilon} := \left\{ \begin{array}{c} (x_1, x_2, x_3, \xi_1, \xi_2, \xi_3, \Gamma) \\ \in W(\mathbf{P}(E)) \end{array} \middle| \begin{array}{c} \xi_1 = \xi_2 = \xi_3, \\ \Gamma \text{ is the net of conics} \\ \text{on the fibre } \mathbf{P}(E(\pi(x_1))) \cong \mathbf{P}_2, \\ \text{containing } \xi_1 \end{array} \right\},$$

and $\tau \in A^1(W(\mathbf{P}(E)))$ the class of

$$\widetilde{\tau} := \left\{ \begin{array}{c} (x_1, x_2, x_3, \xi_1, \xi_2, \xi_3, \Gamma) \\ \in W(\mathbf{P}(E)) \end{array} \middle| \begin{array}{c} x_1 = x_2 = x_3, \\ \Gamma \text{ is the net of conics,} \\ \text{on the fibre } \mathbf{P}(E(\pi(x_1))) \cong \mathbf{P}_2, \\ \text{having a singular point at } x_1 \end{array} \right\}.$$

By [Collino-Fulton (1)] we have :

Lemma 4.3.6.

(1) $\qquad \tau = \epsilon + a + b + c + \mu_1 - \alpha - \beta - \gamma,$

(2) $\quad a^3 = \mu_1 a^2 - \mu_2 a + \mu_3 \quad$ *(and similarly for b and c)*,

(3) $\quad \alpha^3 = 2\mu_1 \alpha^2 - (\mu_1^2 + \mu_2)\alpha + \mu_1 \mu_2 - \mu_3 \quad$ *(and similarly for β and γ)*,

(4) $\quad a\beta = a^2 + \beta^2 - \mu_1 \beta + \mu_2$

$\qquad$ *(and similarly for a, γ; b, α; b, γ; c, α; c, β respectively)*,

(5) $\quad \epsilon\alpha = \epsilon\beta = \epsilon\gamma$,

(6) $\quad \tau a = \tau b = \tau c$,

(7) $\quad \epsilon\tau = 0$.

Now we want to describe the classes

$$\widehat{\Phi}^*(a), \widehat{\Phi}^*(b), \widehat{\Phi}^*(c), \widehat{\Phi}^*(\alpha), \widehat{\Phi}^*(\beta), \widehat{\Phi}^*(\gamma), \widehat{\Phi}^*(\epsilon), \widehat{\Phi}^*(\tau) \in A^1(\widehat{H}^3(\mathbf{P}(E)/X)).$$

Let

$$p_1, p_2, p_3 : \widehat{H}^3(\mathbf{P}(E)/X) \longrightarrow \mathbf{P}(E),$$

$$q_1, q_2, q_3 : \widehat{H}^3(\mathbf{P}(E)/X) \longrightarrow \mathrm{Hilb}^2(\mathbf{P}(E)/X),$$

$$q : \widehat{H}^3(\mathbf{P}(E)/X) \longrightarrow \mathrm{Hilb}^3(\mathbf{P}(E)/X)$$

be the projections.

Remark 4.3.7.

$$\widehat{\Phi}^*(a) = p_1^*(c_1(\mathcal{O}_{\mathbf{P}(E)}(1))), \quad \widehat{\Phi}^*(b) = p_2^*(c_1(\mathcal{O}_{\mathbf{P}(E)}(1))), \quad \widehat{\Phi}^*(c) = p_3^*(c_1(\mathcal{O}_{\mathbf{P}(E)}(1))),$$

$$\widehat{\Phi}^*(\alpha) = q_1^* axe^*(c_1(T_{2,E}^*)), \quad \widehat{\Phi}^*(\beta) = q_2^* axe^*(c_1(T_{2,E}^*)), \quad \widehat{\Phi}^*(\gamma) = q_3^* axe^*(c_1(T_{2,E}^*)).$$

Let $\overline{A} \in A^1(\mathrm{Hilb}^3(\mathbf{P}(E)/X))$ be the class of $Al^3(\mathbf{P}(E)/X)$. Then we have $\widehat{\Phi}^*(\epsilon) = q^*(\overline{A})$.

$\widehat{\Phi}^*(\tau)$ is the class of the subvariety

$$\left\{ \begin{array}{c} (x_1, x_2, x_3, Z_1, Z_2, Z_3, Z) \\ \in \widehat{H}^3(\mathbf{P}(E)/X) \end{array} \;\middle|\; \begin{array}{c} x_1 = x_2 = x_3, \\ \text{and with } F = \mathbf{P}(E(\pi(x_1))) \\ \mathrm{m}_{F,x_1}^2 \text{ is the ideal of } Z \text{ in } \mathcal{O}_{F,x_1} \end{array} \right\}.$$

Proof: The statements on $\widehat{\Phi}^*(a), \widehat{\Phi}^*(b), \widehat{\Phi}^*(c), \widehat{\Phi}^*(\alpha), \widehat{\Phi}^*(\beta), \widehat{\Phi}^*(\gamma)$ follow easily from the definitions. By definition $\widehat{\Phi}^*(\epsilon)$ is the class of the subvariety

$$\left\{ \begin{array}{c} (x_1, x_2, x_3, Z_1, Z_2, Z_3, Z) \\ \in \widehat{H}^3(\mathbf{P}(E)/X) \end{array} \;\middle|\; \begin{array}{c} \text{the lines } axe(Z_1), axe(Z_2), axe(Z_3) \\ \text{through } Z_1, Z_2, Z_3 \text{ in the} \\ \text{fibre } F = \mathbf{P}(E(\pi(x_1))) \cong \mathbf{P}_2 \\ \text{are equal and } \phi_{2,3}(Z) \\ \text{is the net of conics in } F, \\ \text{containing the line } axe(Z_1). \end{array} \right\}.$$

We consider this condition fibrewise. As $\phi_{2,3}(Z)$ is the kernel of the restriction map $r_Z : H^0(\mathbf{P}_2, \mathcal{O}_{\mathbf{P}_2}(2)) \longrightarrow H^0(Z, \mathcal{O}_Z \otimes \mathcal{O}_{\mathbf{P}_2}(2))$, the condition on $\phi_{2,3}(Z)$, means that Z is a subscheme of the line $axe(Z_1)$ through Z_1. So also Z_2 and Z_3 are subschemes of $axe(Z_1)$, and the conditions on $axe(Z_2)$ and $axe(Z_3)$ are fulfilled automatically. So we get $\widehat{\Phi}^*(\epsilon) = q^*(\overline{A})$.

By definition $\widehat{\Phi}^*(\tau)$ is the class of the subvariety

$$\left\{ \begin{array}{c} (x_1, x_2, x_3, Z_1, Z_2, Z_3, Z) \\ \in \widehat{H}^3(\mathbf{P}(E)/X) \end{array} \right| \left. \begin{array}{c} x_1 = x_2 = x_3 \\ \text{and } \phi_{2,3}(Z) \text{ is the net of conics} \\ \text{in the fibre } \mathbf{P}(E(\pi(x_1))) = \mathbf{P}_2, \\ \text{having a singular point at } x_1 \end{array} \right\}$$

Let $(x_1, x_2, x_3, Z_1, Z_2, Z_3, Z)$ be a point of this subvariety. The condition on $\phi_{2,3}(Z)$ means that Z lies in the subscheme $\widetilde{Z} \subset F = \mathbf{P}(E(\pi(x_1)))$ with support x_1 which is defined by $\mathbf{m}_{F,x_1}^2$ in $\mathcal{O}_{F,x_1}$. $\widetilde{Z}$ is a subscheme of length 3 of $\mathbf{P}(E(\pi(x_1)))$, so we have $Z = \widetilde{Z}$. As x_1, x_2, x_3 are subschemes of Z, the condition $x_1 = x_2 = x_3$ follows automatically from the condition on $\phi_{2,3}(Z)$. The result follows.   □

Now we turn to the variety $\widetilde{\mathrm{Hilb}}^3(\mathbf{P}(E)/X)$ of triangles in the fibres of $\mathbf{P}(E)$ with a marked side. Via $res : \widetilde{\mathrm{Hilb}}^3(\mathbf{P}(E)/X) \longrightarrow \mathbf{P}(E)$ we regard $\widetilde{\mathrm{Hilb}}^3(\mathbf{P}(E)/X)$ as a subscheme of $\mathbf{P}(E) \times_X \mathrm{Hilb}^2(\mathbf{P}(E)/X) \times_X \mathrm{Hilb}^3(\mathbf{P}(E)/X)$:

$$\widetilde{\mathrm{Hilb}}^3(\mathbf{P}(E)/X) = \left\{ (x, Z_1, Z) \mid x \subset Z_1 \subset Z, \; res(Z_1, Z) = x \right\}.$$

So we have a natural morphism

$$\pi_{147} : \widehat{H}^3(\mathbf{P}(E)/X) \longrightarrow \widetilde{\mathrm{Hilb}}^3(\mathbf{P}(E)/X);$$
$$(x_1, x_2, x_3, Z_1, Z_2, Z_3, Z) \longmapsto (x_1, Z_1, Z)$$

Let

$$\widehat{\pi} := \pi_{147} \circ \widehat{\Phi}^{-1} : W(\mathbf{P}(E)) \longrightarrow \widetilde{\mathrm{Hilb}}^3(\mathbf{P}(E)/X).$$

Let

$$\widetilde{p}_1 : \widetilde{\mathrm{Hilb}}^3(\mathbf{P}(E)/X) \longrightarrow \mathbf{P}(E),$$
$$\widetilde{p}_2 : \widetilde{\mathrm{Hilb}}^3(\mathbf{P}(E)/X) \longrightarrow \mathrm{Hilb}^2(\mathbf{P}(E)/X),$$
$$\widetilde{\pi} : \widetilde{\mathrm{Hilb}}^3(\mathbf{P}(E)/X) \longrightarrow \mathrm{Hilb}^3(\mathbf{P}(E)/X)$$

be the projections. Let $\widetilde{\phi}_2 : \mathrm{Hilb}^2(\mathbf{P}(E)/X) \longrightarrow \mathbf{P}(\mathrm{Sym}^2(T_{2,E}))$ be the isomorphism from lemma 4.1.15 with $\widetilde{\phi}_2^*(\mathcal{O}_{\mathbf{P}(\mathrm{Sym}^2(T_{2,E}))}(1)) = (q_2)_* p^*(\mathcal{O}_{\mathbf{P}(E)}(1))$. Here

$$Z_2(\mathbf{P}(E)/X)$$

$$\swarrow^{p} \qquad\qquad \searrow^{q_2}$$

$$\mathbf{P}(E) \qquad\qquad\qquad \mathrm{Hilb}^2(\mathbf{P}(E)/X)$$

are the natural projections of the universal subscheme.

Definition 4.3.8. We put

$$H := \widetilde{p}_1^*(c_1(\mathcal{O}_{\mathbf{P}(E)}(1))),$$
$$H_2 := \widetilde{p}_2^*(q_2)_* p^* c_1(\mathcal{O}_{\mathbf{P}(E)}(1)) = \widetilde{p}_2^* \widetilde{\phi}_2^* c_1(\mathcal{O}_{\mathbf{P}(\mathrm{Sym}^2(T_{2,E}))}(1)),$$
$$P := \widetilde{p}_2^* a x e^*(c_1(T_{2,E}^*)),$$
$$A = \widetilde{\pi}^*(\overline{A}).$$

We want to show that H, H_2, P, A generate $A^*(\widetilde{\mathrm{Hilb}}^3(\mathbf{P}(E)/X))$ as an $A^*(X)$-algebra and to determine the relations. For this we first determine the classes $\widehat{\pi}^*(H), \widehat{\pi}^*(H_2), \widehat{\pi}^*(P), \widehat{\pi}^*(A) \in A^1(W(\mathbf{P}(E)))$.

Lemma 4.3.9. $\widehat{\pi}^*(H) = a, \widehat{\pi}^*(H_2) = b + c, \widehat{\pi}^*(P) = \alpha, \widehat{\pi}^*(A) = \epsilon.$

Proof: $\widehat{\pi}^*(H) = a, \widehat{\pi}^*(P) = \alpha, \widehat{\pi}^*(A) = \epsilon$ follow immediately from the definitions and remark 4.3.7. Now we show $\widehat{\pi}^*(H_2) = b + c$. Let $F(E) \subset \mathbf{P}(E) \times_X \check{\mathbf{P}}(E)$ be the incidence variety

$$F(E) := \Big\{ (x, l) \in \mathbf{P}(E) \times_X \check{\mathbf{P}}(E) \mid x \in l \Big\}$$

and

$$F(E)$$
$$\swarrow^{p_1} \qquad \searrow^{p_2}$$
$$\mathbf{P}(E) \qquad\qquad \check{\mathbf{P}}(E)$$

the projections. It is easy to see that there is an isomorphism $\overline{\psi} : F(E) \longrightarrow \mathbf{P}(T_{2,E})$ over $\check{\mathbf{P}}(E)$ with $\overline{\psi}^*(\mathcal{O}_{\mathbf{P}(T_{2,E})}(1)) = p_1^*(\mathcal{O}_{\mathbf{P}(E)}(1))$. Let

$$r_{2,3} : W(\mathbf{P}(E)) \longrightarrow (\mathbf{P}(E) \times_X \check{\mathbf{P}}(E)) \times_X (\mathbf{P}(E) \times_X \check{\mathbf{P}}(E));$$
$$(x_1, x_2, x_3, \xi_1, \xi_2, \xi_3, \Gamma) \longmapsto ((x_2, \xi_2), (x_3, \xi_3)).$$

We see from the definitions that the image $r_{2,3}(W(\mathbf{P}(E)))$ lies in the subvariety $F(E) \times_X F(E)$ of $(\mathbf{P}(E) \times_X \check{\mathbf{P}}(E)) \times_X (\mathbf{P}(E) \times_X \check{\mathbf{P}}(E))$. The diagram

$$
\begin{array}{ccccc}
W(\mathbf{P}(E)) & \xrightarrow{\;r_{2,3}\;} & F(E) \times_X F(E) & \xrightarrow{\;\overline{\psi} \times \overline{\psi}\;} & \mathbf{P}(T_{2,E}) \times_X \mathbf{P}(T_{2,E}) \\
\downarrow{\scriptstyle \widehat{\pi}} & & & & \downarrow{\scriptstyle \eta} \\
\widetilde{\mathrm{Hilb}}^3(\mathbf{P}(E)/X) & \xrightarrow{\;\widetilde{p}_2\;} & \mathrm{Hilb}^2(\mathbf{P}(E)/X) & \xrightarrow{\;\widetilde{\phi}_2\;} & \mathbf{P}(\mathrm{Sym}^2(T_{2,E}))
\end{array}
$$

commutes. Here η is the morphism defined by the natural map $T_{2,E} \otimes T_{2,E} \longrightarrow$ $\mathrm{Sym}^2(T_{2,E})$. With respect to the projections

$$r_1, r_2 : \mathbf{P}(T_{2,E}) \times_{\dot{\mathbf{P}}(E)} \mathbf{P}(T_{2,E}) \longrightarrow \mathbf{P}(T_{2,E})$$

we have:

$$\eta^*(c_1(\mathcal{O}_{\mathbf{P}(\mathrm{Sym}^2(T_{2,E}))}(1))) = r_1^*(c_1(\mathcal{O}_{\mathbf{P}(T_{2,E})}(1))) + r_2^*(c_1(\mathcal{O}_{\mathbf{P}(T_{2,E})}(1))).$$

By $\overline{\psi}^*(\mathcal{O}_{\mathbf{P}(T_{2,E})}(1)) = p_1^*(\mathcal{O}_{\mathbf{P}(E)}(1))$ the result follows.   □

Now we can give a first description of the Chow ring $A^*(\widetilde{\mathrm{Hilb}}^3(\mathbf{P}(E)/X))$.

Proposition 4.3.10. $\widehat{\pi}^* : A^*(\widetilde{\mathrm{Hilb}}^3(\mathbf{P}(E)/X)) \longrightarrow A^*(W(\mathbf{P}(E)))$ *is injective.* $\widehat{\pi}^*(A^*(\widetilde{\mathrm{Hilb}}^3(\mathbf{P}(E)/X)))$ *is the $A^*(X)$-subalgebra of $A^*(W(\mathbf{P}(E)))$ generated by*

$$\widehat{\pi}^*(H) = a, \ \widehat{\pi}^*(H_2) = b + c, \ \widehat{\pi}^*(P) = \alpha, \ \widehat{\pi}^*(A) = \epsilon.$$

Proof: The classes which we called A, H, H_2, P in section 4.2 will now be called $A_{\mathbf{P}_2}$, $H_{\mathbf{P}_2}, H_{2,\mathbf{P}_2}, P_{\mathbf{P}_2}$. We see that the restrictions of A, H, H_2, P to a fibre $\widetilde{\mathrm{Hilb}}^3(\mathbf{P}_2)$ are $A_{\mathbf{P}_2}, H_{\mathbf{P}_2}, H_{2,\mathbf{P}_2}, P_{\mathbf{P}_2}$. Then by the theorem of Leray-Hirsch for the Chow groups [Collino-Fulton (1)] the monomials in A, H, P, H_2 occuring in the intersection tables at the end of section 4.2 form a basis of $A^*(\widetilde{\mathrm{Hilb}}^3(\mathbf{P}(E)/X))$ as a free $A^*(X)$-module (as $\widetilde{\mathrm{Hilb}}^3(\mathbf{P}_2)$ has a cell decomposition). So we only have to see that $\widehat{\pi}^*$ is injective. Let $\widehat{\pi}_{\mathbf{P}_2} : \widehat{H}^3(\mathbf{P}_2) \longrightarrow \widetilde{\mathrm{Hilb}}^3(\mathbf{P}_2)$ be the restriction of $\widehat{\pi}$ to a fibre $\widehat{H}^3(\mathbf{P}_2)$. The orientation classes $[*]$ of $\widetilde{\mathrm{Hilb}}^3(\mathbf{P}_2)$ and $[**]$ of $\widehat{H}^3(\mathbf{P}_2)$ fulfill $\widehat{\pi}_{\mathbf{P}_2}^*([*]) = 3[**]$, $\widehat{\pi}_*([**]) = [*]$, as $\widehat{\pi}_{\mathbf{P}_2}$ is generically finite of degree 3. As both $\widetilde{\mathrm{Hilb}}^3(\mathbf{P}_2)$ and $\widehat{H}^3(\mathbf{P}_2)$ have a cell decomposition, the intersection product in complementary dimensions gives a nondegenerate pairing of free $\mathbf{Z}$-modules for both varieties. So $\widehat{\pi}_{\mathbf{P}_2}^*$ is injective. As a homomorphism of free $A^*(X)$-modules

$$\widehat{\pi}^* = \widehat{\pi}_{\mathbf{P}_2}^* \otimes 1_{A^*(X)} : A^*(\widetilde{\mathrm{Hilb}}^3(\mathbf{P}(E)/X))$$
$$= A^*(\widetilde{\mathrm{Hilb}}^3(\mathbf{P}_2)) \otimes A^*(X) \longrightarrow A^*(\widehat{H}^3(\mathbf{P}_2)) \otimes A^*(X) = A^*(\widehat{H}^3(\mathbf{P}(E)/X))$$

is one to one. So $\widehat{\pi}^*$ is injective.   □

We now describe $A^*(\widetilde{\mathrm{Hilb}}^3(\mathbf{P}(E)/X))$ directly by generators and relations.

Theorem 4.3.11.

$$A^*(\widetilde{\mathrm{Hilb}}^3(\mathbf{P}(E)/X)) = \frac{A^*(X)[H_2, P, H, A]}{(I_1, I_2, I_3, I_4, I_5, I_6)}$$

where

$$I_1 := H^3 - \mu_1 H^2 + \mu_2 H - \mu_3,$$
$$I_2 := P(P - \mu_1)^2 + \mu_2(P - \mu_1) + \mu_3,$$
$$I_3 := H_2^3 - 3PH_2^2 + H_2(6P^2 - 4P\mu_1 + 4\mu_2) - 4(P^3 - P^2\mu_1 + P\mu_2),$$
$$I_4 := A(H^2 - PH + P(P - \mu_1) + \mu_2),$$
$$I_5 := A(A - 3P + H + H_2 + \mu_1),$$
$$I_6 := -AH_2^2 + \mu_1(-H_2^2 + H_2 P + 2HH_2 - 2HP) + H_2^2 P - H_2 P^2 + HH_2^2$$
$$\qquad - 3HH_2 P + 2HP^2 - H^2 H_2 + 2H^2 P + A(H_2 P + 2HH_2 - 2HP).$$

Proof: We have $\widetilde{p}_1^*(A^*(\mathbf{P}(E))) = A^*(X)[H]/(I_1)$. Furthermore

$$P = \widetilde{p}_2^* axe^*(c_1(Q_{1,E})) + \mu_1$$

and thus

$$\widetilde{p}_2^* axe^*(A^*(\check{\mathbf{P}}(E))) = \frac{A^*(X)[P]}{((P - \mu_1)^3 + \mu_1(P - \mu_1)^2 + \mu_2(P - \mu_1) + \mu_3)}$$
$$= A^*(X)[P]/(I_2).$$

We have

$$\widetilde{p}_2^* axe^* c_1(T_{2,E}) = -P,$$
$$\widetilde{p}_2^* axe^* c_2(T_{2,E}) = P(P - \mu_1) + \mu_2.$$

So we get by 3.1.9

$$\widetilde{p}_2^* axe^* c(\mathrm{Sym}^2(T_{2,E})) = 1 - 3P + (6P^2 - 4P\mu_1 + 4\mu_2) - 4(P^3 - P^2\mu_1 + P\mu_2)$$

and thus

Remark 4.3.12. The $A^*(X)$-subalgebra of $A^*(\mathbf{P}(E)/X))$ generated by H, P, H_2 is

$$(\widetilde{p}_1 \times \widetilde{p}_2)^*(A^*(\mathbf{P}(E) \times_X \mathrm{Hilb}^2(\mathbf{P}(E)/X))) = \frac{A^*(X)[H, H_2, P]}{(I_1, I_2, I_3)}.$$

Let $\widetilde{A} \subset \widetilde{\mathrm{Hilb}}^3(\mathbf{P}(E)/X)$ be the subvariety defined by

$$\widetilde{A} := \left\{ \begin{array}{c} (x, Z_1, Z) \\ \in \widetilde{\mathrm{Hilb}}^3(\mathbf{P}(E)/X) \end{array} \middle| \begin{array}{c} Z \text{ lies on a line} \\ \text{in the fibre } \mathbf{P}(E(\pi(x))) \\ \text{passing through } x \end{array} \right\}.$$

Via

$$axe.\widetilde{p}_2 = axe.\widetilde{\pi} : \widetilde{A} \longrightarrow \check{\mathbf{P}}(E)$$

$\widetilde{A}$ is a variety over $\check{\mathbf{P}}(E)$.

$$\widetilde{p}_1 \times \widetilde{\pi} : \widetilde{\mathrm{Hilb}}^3(\mathbf{P}(E)/X) \longrightarrow \mathbf{P}(E) \times_X \mathrm{Hilb}^3(\mathbf{P}(E)/X)$$

maps $\widetilde{A}$ isomorphically onto $Z_3^{al}(\mathbf{P}(E)/X)$. By lemma 4.1.15 there is an isomorphism

$$\widetilde{\phi} : Z_3^{al}(\mathbf{P}(E)/X) \longrightarrow \mathbf{P}(T_{2,E}) \times_{\check{\mathbf{P}}(E)} \mathbf{P}(\mathrm{Sym}^2(T_{2,E}))$$

over $\check{\mathbf{P}}(E)$ satisfying

$$(\widetilde{p}_1 \times \widetilde{\pi}|_{\widetilde{A}})^* \widetilde{\phi}^*(c_1(\mathcal{O}_{\mathbf{P}(T_{2,E})}(1))) = H|_{\widetilde{A}},$$
$$(\widetilde{p}_1 \times \widetilde{\pi}|_{\widetilde{A}})^* \widetilde{\phi}^*(c_1(\mathcal{O}_{\mathbf{P}(\mathrm{Sym}^2(T_{2,E}))}(1))) = H_2|_{\widetilde{A}}.$$

So we get

$$A^*(\widetilde{A}) = \frac{A^*(X)[H, P, H_2]}{(I_2, I_3, H^2 - PH + P(P - \mu_1) - \mu_2)}.$$

The relation $I_4 = 0$ in $A^*(\widetilde{\mathrm{Hilb}}^3(\mathbf{P}(E)/X))$ follows by $[\widetilde{A}] = A$.

In order to prove the relations $I_5 = 0$, $I_6 = 0$, we want to compute in $A^*(W(\mathbf{P}(E)))$ and use the reations of Collino and Fulton from lemma 4.3.6. The proof of $I_5 = 0$ is simple.

$$\widetilde{\pi}^*(A(A - 3P + H + H_2 + \mu_1)) = \epsilon(\epsilon - 3\alpha + a + b + c + \mu_1)$$
$$= \epsilon(\epsilon - \alpha - \beta - \gamma + a + b + c + \mu_1)$$
$$= \epsilon\tau$$
$$= 0.$$

So $I_5 = 0$ holds. In order to proof $I_6 = 0$, we write the relations in such a way that they can be applied formally (by substituting).

Remark 4.3.13. In $A^*(W(\mathbf{P}(E)))$ the following relations hold:

(1) $a^3 = a^2 \mu_1 - a\mu_2 + \mu_3$ and similarly for b and c,

(2) $\alpha^3 = 2\mu_1\alpha^2 - (\mu_1^2 + \mu_2)\alpha + \mu_1\mu_2 - \mu_3$ and similarly for β and γ,

(3) $\alpha^2 = -b^2 + b\alpha + \mu_1\alpha - \mu_2,$

(4) $\beta^2 = -a^2 + a\beta + \mu_1\beta - \mu_2,$

(5) $\gamma^2 = -a^2 + a\gamma + \mu_1\gamma - \mu_2,$

(6) $\alpha c = -b^2 + b\alpha + c^2,$

(7) $\beta c = -a^2 + a\beta + c^2,$

(8) $\gamma b = -a^2 + a\gamma + b^2,$

(9) $\qquad \epsilon\beta = \epsilon\alpha,$

$\qquad\qquad \epsilon\gamma = \epsilon\alpha,$

(10) $\qquad eb = \epsilon a + (a - b)(c + \mu_1 - \alpha - \beta),$

$\qquad\qquad \epsilon c = \epsilon a + (a - c)(b + \mu_1 - \alpha - \gamma).$

Now we just apply these relations formally. We get

$$0 = \widehat{\pi}^*(A(H^2 - HP + P(P - \mu_1) + \mu_2))$$
$$= \epsilon(a^2 - a\alpha + \alpha(\alpha - \mu_1) + \mu_2)$$
$$= -a^2c - a^2\mu_1 + a^2\alpha + a^2\beta + ac^2 + a\mu_2 - a\alpha\beta$$
$$\quad + b^2c + b^2\mu_1 - b^2\alpha - b^2\beta - bc^2 - b\mu_2 + b\alpha\beta.$$

Furthermore we get

$$\widehat{\pi}^*(H_2^2 P) = (b + c)^2\alpha$$
$$= -b^2c - 3b^2\mu_1 + 4b^2\alpha + 3bc^2 + 3b\mu_2 + c^2\mu_1 - c\mu_2 - 2\mu_3,$$
$$\widehat{\pi}^*(H_2 P^2) = \alpha^2(b + c)$$
$$= -b^2c - 3b^2\mu_1 + 2b^2\alpha + bc^2 + 2b\alpha\mu_1 + b\mu_2 + c^2\mu_1 - c\mu_2 - 2\mu_3,$$
$$\widehat{\pi}^*(HH_2^2) = a(b^2 + 2bc + c^2),$$
$$\widehat{\pi}^*(HH_2 P) = a\alpha(b + c)$$
$$= a(-b^2 + 2b\alpha + c^2),$$
$$\widehat{\pi}^*(HP^2) = a\alpha^2$$
$$= a(-b^2 + b\alpha + \mu_1\alpha - \mu_2),$$
$$\widehat{\pi}^*(H^2 H_2) = a^2(b + c),$$
$$\widehat{\pi}^*(H^2 P) = a^2\alpha,$$
$$\widehat{\pi}^*(AHP) = a\alpha\epsilon,$$
$$\widehat{\pi}^*(AHH_2) = a(ab + ac + 2a\mu_1 - 2a\alpha - a\beta - a\gamma + 2a\epsilon$$
$$\quad - b^2 - 2bc - b\mu_1 + 2b\alpha + b\beta + c^2 - c\mu_1 + c\gamma),$$
$$\widehat{\pi}^*(AH_2 P) = \epsilon\alpha(b + c)$$
$$= a^2b - a^2c + a^2\beta - a^2\gamma + 2ac^2 + 2a\mu_2 - 2a\alpha\beta$$
$$\quad + 2a\alpha\epsilon - b^2\beta - 2bc^2 - b\mu_2 + 2b\alpha\beta + c^2\gamma - c\mu_2,$$
$$\widehat{\pi}^*(AH_2^2) = \epsilon(b + c)^2$$
$$= a^2b + a^2c + 6a^2\mu_1 - 4a^2\alpha - 3a^2\beta - 3a^2\gamma + 4a^2\epsilon - 2abc$$
$$\quad + 2ab\beta + 2ac\gamma - 2a\mu_2 - 2b^2c - 4b^2\mu_1 + 4b^2\alpha + b^2\beta$$
$$\quad + 2bc^2 - 2bc\mu_1 + 3b\mu_2 + c^2\gamma - c\mu_2,$$

$$\widehat{\pi}^*(H_2^2\mu_1) = \mu_1(b^2 + 2bc + c^2),$$
$$\widehat{\pi}^*(H_2 P\mu_1) = \mu_1(-b^2 + 2b\alpha + c^2),$$
$$\widehat{\pi}^*(HH_2\mu_1) = a\mu_1(b + c),$$
$$\widehat{\pi}^*(HP\mu_1) = a\mu_1\alpha.$$

Thus we have

$$\widehat{\pi}^*(-AH_2^2 + \mu_1(-H_2^2 + H_2 P + 2HH_2 - 2HP) + H_2^2 P - H_2 P^2 + HH_2^2$$
$$- 3HH_2 P + 2HP^2 - H^2 H_2 + 2H^2 P + A(H_2 P + 2HH_2 - 2HP))$$
$$= 2(-a^2 c - a^2\mu_1 + a^2\alpha + a^2\beta + ac^2 + a\mu_2 - a\alpha\beta$$
$$+ b^2 c + b^2\mu_1 - b^2\alpha - b^2\beta - bc^2 - b\mu_2 + b\alpha\beta)$$
$$= 0.$$

As $\widehat{\pi}^*$ is injective, the relation $I_6 = 0$ holds in $A^*(\widetilde{\mathrm{Hilb}}^3(\mathbf{P}(E)/X))$.

End of the proof of theorem 4.3.11

The monomials in A, H, H_2, P occuring in the intersection tables at the end of 4.2 form a basis of $\widetilde{\mathrm{Hilb}}^3(\mathbf{P}(E)/X))$ as a free $A^*(X)$-module. On the other hand using the relations $I_1, \ldots, I_6$ we can express any monomial M in A, H, P, H_2 as an $A^*(X)$-linear combination of monomials of the form $A^a H^h P^p H_2^{h_2}$ with

$$h \le 2, \ p \le 2, \ h_2 \le 2, \ a \le 1, \ h + a \le 2, a + h_2 \le 2,$$

i.e. as a linear combination of these monomials. The result follows.   □

In the rest of this section we look at an important special case of $\widetilde{\mathrm{Hilb}}^3(\mathbf{P}(E)/X))$. We put $G := \mathit{Grass}(d - 2, d + 1)$ and let $T := T_{3,d+1}$ be the tautological bundle over G.

Definition 4.3.14. Let $\widetilde{\mathrm{Cop}}^3(\mathbf{P}_d) \subset \widetilde{\mathrm{Hilb}}^3(\mathbf{P}_d) \times G$ be the subvariety

$$\widetilde{\mathrm{Cop}}^3(\mathbf{P}_d) := \left\{ ((Z_1, Z), E) \in \widetilde{\mathrm{Hilb}}^3(\mathbf{P}_d) \times G \ \Big| \ Z \subset E \right\}$$
$$= \left\{ (Z_1, Z, E) \in \mathbf{P}_d^{[2]} \times \mathbf{P}_d^{[3]} \times G \ \Big| \ Z_1 \subset Z \subset E \right\}.$$

Let $F \subset \mathbf{P}_d \times G$ be the incidence variety $F := \left\{ (x, E) \in \mathbf{P}_d \times G \ \big| \ x \subset E \right\}$ with ptojections

$$F$$
$$\swarrow {\scriptstyle p_1} \qquad \searrow {\scriptstyle p_2}$$
$$\mathbf{P}_d \qquad\qquad\qquad G.$$

There is an isomorphism $\phi : F \longrightarrow \mathbf{P}(T)$ over G with $\phi^*(\mathcal{O}_{\mathbf{P}(T)}(1)) = p_1^*(\mathcal{O}_{\mathbf{P}_d}(1))$. We see immediately from the definitions that $\widetilde{\text{Cop}}^3(\mathbf{P}_d)$ is the sub-variety $\widetilde{\text{Hilb}}^3(F/G) \subset \widetilde{\text{Hilb}}^3(\mathbf{P}_d) \times G$. So we get an isomorphism

$$\tilde{\phi} : \widetilde{\text{Cop}}^3(\mathbf{P}_d) \longrightarrow \widetilde{\text{Hilb}}^3(\mathbf{P}(T)/G).$$

The projection $\check{p}_1 : \widetilde{\text{Cop}}^3(\mathbf{P}_d) \longrightarrow \widetilde{\text{Hilb}}^3(\mathbf{P}_d)$ is a birational morphism (a general subscheme of length 3 lies on exactly one plane). It is an isomorphism outside

$$\tilde{A}_0 := \left\{ (Z_1, Z) \in \widetilde{\text{Hilb}}^3(\mathbf{P}_d) \,\middle|\, Z \text{ lies on a line } \right\}.$$

Over a point $(Z_1, Z) \in \tilde{A}_0$, lying on a line l its fibre is

$$\check{p}_1^{-1}(Z_1, Z) = \left\{ E \in G \,\middle|\, E \supset l \right\} = \mathbf{P}_{d-2}$$

The exceptional locus of $\check{p}_1$ is

$$\tilde{A} \cong \mathbf{P}(T_{2,T}) \times_{\check{\mathbf{P}}(T)} \mathbf{P}(\text{Sym}^2(T_{2,T})),$$

in particular it is an irreducible divisor. So we get:

Remark 4.3.15. $\widetilde{\text{Cop}}^3(\mathbf{P}_d)$ is obtained by blowing up $\widetilde{\text{Hilb}}^3(\mathbf{P}_d)$ along $Z_3^{al}(\mathbf{P}_d)$.

Definition 4.3.16. Let $A', H', H_2', P', \mu_1', \mu_2', \mu_3' \in A^*(\widetilde{\text{Cop}}^3(\mathbf{P}_d))$ be the classes

$$A' := \left[\left\{ (E, (Z_1, Z)) \in \widetilde{\text{Cop}}^3(\mathbf{P}_d) \,\middle|\, Z \text{ lies on a line } \right\} \right],$$

$$H' := \left[\left\{ (E, (Z_1, Z)) \in \widetilde{\text{Cop}}^3(\mathbf{P}_d) \,\middle|\, res(Z_1, Z) \text{ lies on a fixed hyperplane } \right\} \right],$$

$$H_2' := \left[\left\{ (E, (Z_1, Z)) \in \widetilde{\text{Cop}}^3(\mathbf{P}_d) \,\middle|\, supp(Z_1) \text{ intersects a fixed hyperplane } \right\} \right],$$

$$P' := \left[\left\{ \begin{matrix} (E, (Z_1, Z)) \\ \in \widetilde{\text{Cop}}^3(\mathbf{P}_d) \end{matrix} \,\middle|\, \begin{matrix} \text{the line passing through } Z_1 \text{ intersects a fixed} \\ \text{2-codimensional linear subspace} \end{matrix} \right\} \right],$$

$$\mu_1' := \left[\left\{ (E, (Z_1, Z)) \in \widetilde{\text{Cop}}^3(\mathbf{P}_d) \,\middle|\, \begin{matrix} E \text{ intersects a fixed 3-codimensional} \\ \text{linear subspace} \end{matrix} \right\} \right],$$

$$\mu_2' := \left[\left\{ (E, (Z_1, Z)) \in \widetilde{\text{Cop}}^3(\mathbf{P}_d) \,\middle|\, \begin{matrix} E \text{ has a one-dimensional intersection} \\ \text{with a fixed 2-codimensional} \\ \text{linear subspace} \end{matrix} \right\} \right],$$

$$\mu_3' := \left[\left\{ (E, (Z_1, Z)) \in \widetilde{\text{Cop}}^3(\mathbf{P}_d) \,\middle|\, E \text{ lies on a fixed hyperplane } \right\} \right].$$

Then we see easily from the definitions :

Remark 4.3.17.

$$\widetilde{\phi}^*(A) = A', \ \widetilde{\phi}^*(H) = H', \ \widetilde{\phi}^*(H_2) = H_2', \ \widetilde{\phi}^*(P) = P',$$
$$\widetilde{\phi}^*(\mu_1) = \mu_1', \ \widetilde{\phi}^*(\mu_2) = \mu_2', \ \widetilde{\phi}^*(\mu_3) = \mu_3'.$$

So theorem 4.3.12 describes the Chow ring of $\widetilde{\mathrm{Cop}}^3(\mathbf{P}_d)$ in terms of classes determined by the position of subschemes relative to lines and planes in $\mathbf{P}_d$.

4.4. The Chow ring of $\text{Hilb}^3(\mathbf{P}(E)/X)$

As in section 4.3 let E be a vector bundle of rank 3 over a smooth variety X. We want to use the results of the previous section about $A^*(\widetilde{\text{Hilb}}^3(\mathbf{P}(E)/X))$, to compute the Chow ring $A^*(\text{Hilb}^3(\mathbf{P}(E)/X))$ of the relative Hilbert scheme. $\widetilde{\text{Hilb}}^3(\mathbf{P}_2)$ has been defined in [Elencwajg-Le Barz (3)] in order to determine the Chow ring of $\mathbf{P}_2^{[3]}$ by generators and relations. There the following classes are introduced:

Definition 4.4.1. Let $\tilde{\pi} : \widetilde{\text{Hilb}}^3(\mathbf{P}_2) \longrightarrow \mathbf{P}_2^{[3]}$ be the projection. Let

$$\check{H}, \check{A} \in A^1(\mathbf{P}_2^{[3]}),$$
$$\check{h}, \check{p}, \check{a} \in A^2(\mathbf{P}_2^{[3]}),$$
$$\check{\alpha}, \check{\beta} \in A^3(\mathbf{P}_2^{[3]})$$

be the classes defined by

$$\check{H} := \tilde{\pi}_*(H),$$
$$\check{A} := \left[\left\{ Z \in \mathbf{P}_2^{[3]} \,\middle|\, Z \text{ lies on a line} \right\}\right],$$
$$\check{h} := \tilde{\pi}_*(H^2),$$
$$\check{p} := \tilde{\pi}_*(P^2),$$
$$\check{a} := \left[\left\{ Z \in \mathbf{P}_2^{[3]} \,\middle|\, Z \text{ lies on a line passing through a fixed point} \right\}\right],$$
$$\check{\alpha} := \left[\left\{ Z \in \mathbf{P}_2^{[3]} \,\middle|\, Z \text{ lies on a fixed line} \right\}\right],$$
$$\check{\beta} := \tilde{\pi}_*(HP^2).$$

Here $H, P \in A^1(\widetilde{\text{Hilb}}^3(\mathbf{P}_2))$ are the classes from definition 4.2.1.

[Elencwajg-Le Barz (3)] get for instance:

Theorem 4.4.2. [Elencwajg-Le Barz (3)]

(1) $\check{H}, \check{A}, \check{h}, \check{p}, \check{a}, \check{\alpha}, \check{\beta}$ *generate* $A^*(\mathbf{P}_2^{[3]})$ *as a ring.*

(2) *Bases of the free $\mathbb{Z}$-modules* $A^i(\mathbf{P}_2^{[3]})$ *are*

$i = 0$: 1;

$i = 1$: $\check{H}, \check{A}$;

$i = 2$: $\check{H}^2, \check{H}\check{A}, \check{a}, \check{h}, \check{p}$;

$i = 3$: $\check{H}^3, \check{h}\check{H}, \check{H}^2\check{A}, \check{H}\check{a}, \check{\alpha}, \check{\beta}$;

$i = 4$: $\check{H}^2\check{a}, \check{H}\check{\alpha}, \check{H}^2\check{h}, \check{h}^2, \check{h}\check{p}$;

$i = 5$: $\check{H}\check{h}^2, \check{H}\check{h}\check{p}$;

$i = 6$: $\check{h}^3$.

Elencwajg and Le Barz determine all the relations between the generators. We will first define some classes in $A^*(\mathrm{Hilb}^3(\mathbf{P}(E)/X))$ as relative versions of the classes in [Elencwajg-Le Barz (3)].

Definition 4.4.3. Let $\widetilde{\pi} : \widetilde{\mathrm{Hilb}}^3(\mathbf{P}(E)/X) \longrightarrow \mathrm{Hilb}^3(\mathbf{P}(E)/X)$ be the projection. Let

$$\bar{H} := \widetilde{\pi}_*(H) \in A^1(\mathrm{Hilb}^3(\mathbf{P}(E)/X)),$$
$$\bar{h} := \widetilde{\pi}_*(H^2), \quad \bar{p} := \widetilde{\pi}_*(P^2) \in A^2(\mathrm{Hilb}^3(\mathbf{P}(E)/X)),$$
$$\bar{\beta} := \widetilde{\pi}_*(HP^2) \in A^3(\mathrm{Hilb}^3(\mathbf{P}(E)/X)).$$

Here $H, P \in A^1(\widetilde{\mathrm{Hilb}}^3(\mathbf{P}(E)/X)$ are the classes from definition 4.3.8. Let $i : Al^3(\mathbf{P}(E)/X) \longrightarrow \mathrm{Hilb}^3(\mathbf{P}(E)/X)$ be the embedding and

$$axe : Al^3(\mathbf{P}(E)/X) \longrightarrow \check{\mathbf{P}}(E)$$

the axial morphism from 4.1.14. Let again $T_{2,E}$ be the tautological subbundle on $\check{\mathbf{P}}(E)$ and $\widetilde{P} := axe^*(c_1(T_{2,E}^*))$. We put

$$\bar{A} := [Al^3(\mathbf{P}(E)/X)] = i_*(1) \in A^1(\mathrm{Hilb}^3(\mathbf{P}(E)/X)),$$
$$\bar{a} := i_*(\widetilde{P}) \in A^2(\mathrm{Hilb}^3(\mathbf{P}(E)/X)),$$
$$\bar{\alpha} := i_*(\widetilde{P}^2) \in A^3(\mathrm{Hilb}(\mathbf{P}(E)/X)).$$

Proposition 4.4.4.

(1) $\bar{H}, \bar{A}, \bar{h}, \bar{p}, \bar{a}, \bar{\alpha}, \bar{\beta}$ generate $A^*(\mathrm{Hilb}^3(\mathbf{P}(E)/X)))$ as an $A^*(X)$-algebra.

(2) The $A^i(\mathrm{Hilb}^3(\mathbf{P}(E)/X)))$ are free $A^*(X)$-modules with basis

$i = 0$: 1;

$i = 1$: $\bar{H}, \bar{A}$;

$i = 2$: $\bar{H}^2, \bar{H}\bar{A}, \bar{a}, \bar{h}, \bar{p}$;

$i = 3$: $\bar{H}^3, \bar{h}\bar{H}, \bar{H}^2\bar{A}, \bar{H}\bar{a}, \bar{\alpha}, \bar{\beta}$;

$i = 4$: $\bar{H}^2\bar{a}, \bar{H}\bar{\alpha}, \bar{H}^2\bar{h}, \bar{h}^2, \bar{h}\bar{p}$;

$i = 5$: $\bar{H}\bar{h}^2, \bar{H}\bar{h}\bar{p}$,

$i = 6$: $\bar{h}^3$.

Proof: (1) follows from (2). Immediately from the definitions we get for the fibre $F \cong \mathbf{P}_2^{[3]}$ of $\mathrm{Hilb}^3(\mathbf{P}(E)/X)$ over a point $x \in X$:

$$\bar{H}|_F = \breve{H}, \ \bar{A}|_F = \breve{A},$$
$$\bar{h}|_F = \breve{h}, \ \bar{p}|_F = \breve{p}, \ \bar{a}|_F = \breve{a},$$
$$\bar{\alpha}|_F = \breve{\alpha}, \ \bar{\beta}|_F = \breve{\beta}.$$

As $\mathbf{P}_2^{[3]}$ has a cell decomposition, we get (2) from the theorem of Leray-Hirsch for Chow groups [Collino-Fulton (1)] and 4.4.2.   □

In order to be able to compute the image of these classes under $\tilde{\pi}^*$, we prove a result on the relations between $\tilde{\pi}^*, \tilde{\pi}_*, \hat{\pi}^*, \hat{\pi}_*$. Remember that $\hat{\pi}$ is defined by

$$\hat{\pi} : \hat{H}^3(\mathbf{P}(E)/X) \longrightarrow \widetilde{\mathrm{Hilb}}^3(\mathbf{P}(E)/X);$$
$$(x_1, x_2, x_3, Z_1, Z_2, Z_3, Z) \longmapsto (x_1, Z_1, Z).$$

We also consider

$$\hat{\pi}_2 : \hat{H}^3(\mathbf{P}(E)/X) \longrightarrow \widetilde{\mathrm{Hilb}}^3(\mathbf{P}(E)/X);$$
$$(x_1, x_2, x_3, Z_1, Z_2, Z_3, Z) \longrightarrow (x_2, Z_2, Z).$$

Let

$$\tilde{p}_{1,2} : \widetilde{\mathrm{Hilb}}^3(\mathbf{P}(E)/X) \longrightarrow \mathbf{P}(E) \times_X \mathrm{Hilb}^2(\mathbf{P}(E)/X);$$
$$(x, Z_1, Z) \longmapsto (x, Z_1).$$

Lemma 4.4.5. *For* $W \in \tilde{p}_{1,2}^*(A^*(\mathbf{P}(E) \times_X \mathrm{Hilb}^2(\mathbf{P}(E)/X)))$ *we have*

$$\tilde{\pi}^* \tilde{\pi}_*(W) = W + \hat{\pi}_*(\hat{\pi}_2^*(W)).$$

Proof: Let $W = \sum_i a_i [X_i]$ be the representation of W as a linear combination of classes of irreducible varieties. Then we have $\tilde{\pi}^* \tilde{\pi}_*(W) = \sum_i a_i \tilde{\pi}^* \tilde{\pi}_*([X_i])$. So it is enough to show the result for $W = \tilde{p}_{1,2}^*([Y])$, where $Y \subset \mathbf{P}(E) \times_X \mathrm{Hilb}^2(\mathbf{P}(E)/X)$ is an irreducible subvariety. By the definitions we get

$$\tilde{\pi}_* \tilde{p}_{1,2}^*([Y]) = \left[\left\{ Z \in \mathrm{Hilb}^3(\mathbf{P}(E)/X) \ \middle| \ \begin{array}{c} \text{there is a subscheme } Z_1 \subset Z \\ \text{of length 2 with } (res(Z_1, Z), Z_1) \in Y \end{array} \right\} \right].$$

So we also have

$$\widetilde{\pi}^* \widetilde{\pi}_* \widetilde{p}^*_{1,2}([Y])$$

$$= \left[\left\{ \begin{array}{c|c} (x, Z_1, Z) & \text{there is a subscheme } Z'_1 \subset Z \\ \in \widetilde{\text{Hilb}}^3(\mathbf{P}(E)/X) & \text{of length 2 with } (res(Z'_1, Z), Z'_1) \in Y \end{array} \right\} \right]$$

$$= \left[\left\{ (x, Z_1, Z) \in \widetilde{\text{Hilb}}^3(\mathbf{P}(E)/X) \,\middle|\, (res(Z_1, Z), Z_1) \in Y \right\} \right]$$

$$+ \left[\left\{ \begin{array}{c|c} (x, Z_1, Z) & \text{there is a subscheme } Z'_1 \subset Z \\ \in \widetilde{\text{Hilb}}^3(\mathbf{P}(E)/X) & \text{of length 2 with } x \subset Z'_1 \\ & \text{and } (res(Z'_1, Z), Z'_1) \in Y \end{array} \right\} \right]$$

$$= \widetilde{p}^*_{1,2}([Y]) + \widehat{\pi}_*(\widehat{\pi}^*_2([Y])). \qquad \square$$

So we can obtain $A^*(\text{Hilb}^3(\mathbf{P}(E)/X))$ as a subring of $A^*(\widetilde{\text{Hilb}}^3(\mathbf{P}(E)/X))$.

Theorem 4.4.6. $\widetilde{\pi}^* : A^*(\text{Hilb}^3(\mathbf{P}(E)/X)) \longrightarrow A^*(\widetilde{\text{Hilb}}^3(\mathbf{P}(E)/X))$ *is injective, and* $\widetilde{\pi}^*(A^*(\text{Hilb}^3(\mathbf{P}(E)/X)))$ *is the* $A^*(X)$-*subalgebra generated by*

$\widetilde{\pi}^*(\bar{H}) = H + H_2,$

$\widetilde{\pi}^*(\bar{A}) = A,$

$\widetilde{\pi}^*(\bar{h}) = H^2 + H_2 P - 2P^2 + 2P\mu_1 - 2\mu_2,$

$\widetilde{\pi}^*(\bar{p}) = P^2 - HP + HH_2 - H^2 + AH + \mu_1(-P + H_2 + 2H + A) + \mu_1^2 - 2\mu_2,$

$\widetilde{\pi}^*(\bar{a}) = AP,$

$\widetilde{\pi}^*(\bar{\alpha}) = AP^2,$

$\widetilde{\pi}^*(\bar{\beta}) = H(3P^2 - 2H_2 P + H_2^2 + HP - HH_2) + A(P^2 - HP + HH_2)$
$\qquad + \mu_1(2P^2 - 2H_2 P + H_2^2 - HP + HH_2 + H^2 - AP + AH_2 - AH)$
$\qquad + \mu_1^2(-2P + H_2 - H) + \mu_2(-H_2 + H + A) + 2\mu_1\mu_2 + \mu_3.$

Proof: By proposition 4.4.4 $A^*(\text{Hilb}^3(\mathbf{P}(E)/X))$ is as an $A^*(X)$-algebra generated by $\bar{H}, \bar{A}, \bar{h}, \bar{p}, \bar{a}, \bar{\alpha}, \bar{\beta}$. For each fibre $F \cong \mathbf{P}_2$ the map

$$\widetilde{\pi}^*_{\mathbf{P}_2} := \widetilde{\pi}^*|_{F^{[3]}} : A^*(\mathbf{P}_2^{[3]}) \longrightarrow A^*(\widetilde{\text{Hilb}}^3(\mathbf{P}_2))$$

is one to one. As a homomorphism of $A^*(X)$-modules $\widetilde{\pi}^*$ is just

$$\widetilde{\pi}^*_{\mathbf{P}_2} \otimes 1_{A^*(X)} : A^*(\text{Hilb}^3(\mathbf{P}(E)/X)$$
$$= A^*(\mathbf{P}_2^{[3]}) \otimes A^*(X) \longrightarrow A^*(\widetilde{\text{Hilb}}^3(\mathbf{P}_2)) \otimes A^*(X) = A^*(\mathbf{P}(E)/X);$$

so it is one to one.

We still have to determine the images of the generators under $\widetilde{\pi}^*$. By definition 4.3.8 we have $A = \widetilde{\pi}^*(\bar{A})$. A is the class of

$$
\tilde{A} := \left\{ \begin{array}{c} (x, Z_1, Z) \\ \in \widetilde{\mathrm{Hilb}}^3(\mathbf{P}(E)/X) \end{array} \middle| \begin{array}{c} Z \text{ lies on a line} \\ \text{in the fibre } \mathbf{P}(E(\pi(x))) \\ \text{passing through } x \end{array} \right\}.
$$

Let $\pi' := \widehat{\pi}|_{\widetilde{A}}$. Let again $\widetilde{p}_2 : \widetilde{\mathrm{Hilb}}^3(\mathbf{P}(E)/X) \longrightarrow \mathrm{Hilb}^2(\mathbf{P}(E)/X) = Al^2(\mathbf{P}(E)/X)$ be the projection and $p'_2 := \widetilde{p}_2|_{\widetilde{A}}$. Then we have by definition

$$
P = \widetilde{p}_2^* axe^*(c_1(T^*_{2,E})),
$$
$$
\widetilde{P} = axe^*(c_1(T^*_{2,E})) \in A^1(Al^3(\mathbf{P}(E)/X)).
$$

The diagram

$$
\begin{array}{ccc}
 & \tilde{A} & \\
\swarrow{\scriptstyle p'_2} & & \searrow{\scriptstyle \pi'} \\
Al^2(\mathbf{P}(E)/X) & & Al^3(\mathbf{P}(E)/X) \\
\searrow{\scriptstyle axe} & & \swarrow{\scriptstyle axe} \\
 & \check{\mathbf{P}}(E) &
\end{array}
$$

commutes. So we get $(\pi')^*(\widetilde{P}) = P|_{\widetilde{A}}$ and thus $\widetilde{\pi}^*(\bar{a}) = AP$, $\widetilde{\pi}^*(\check{a}) = AP^2$. By lemma 4.4.5, lemma 4.3.9, remark 4.3.13 and the projection formula we have

$$
\begin{aligned}
\widetilde{\pi}^*(\bar{H}) &= H + \widehat{\pi}_*(b) \\
&= H + H_2, \\
\widetilde{\pi}^*(\bar{h}) &= H^2 + \widehat{\pi}_*(b^2) \\
&= H^2 + \widehat{\pi}_*(ba - \alpha^2 + \mu_1\alpha - \mu_2) \\
&= H^2 + H_2 P - P^2 + 2\mu_1 P - 2\mu_2, \\
\widetilde{\pi}^*(\bar{p}) &= P^2 + \frac{1}{2}\widehat{\pi}_*(\beta^2 + \gamma^2) \\
&= P^2 + \frac{1}{2}\widehat{\pi}_*(a(\beta + \gamma) - 2a^2 + \mu_1(\beta + \gamma) - 2\mu_2) \qquad 4.3.13(4),(5) \\
&= P^2 + \frac{1}{2}\widehat{\pi}_*((a + \mu_1)(\epsilon + a + (b + c) - \alpha - \tau + \mu_1) - 2a^2 + 2\mu_2) \qquad 4.3.6(1) \\
&= P^2 - HP + HH_2 - H^2 + AH + \mu_1(-P + H_2 + 2H + A) + \mu_1^2 - 2\mu_2, \\
\widetilde{\pi}^*(\bar{\beta}) &= HP^2 + \frac{1}{2}\widehat{\pi}_*(b\beta^2 + c\gamma^2).
\end{aligned}
$$

Furthermore we have

$$
b\beta^2 + c\gamma^2 = b(\alpha\beta - a^2 + \mu_1\beta - \mu_2) + c(a\gamma - a^2 + \mu_1\gamma - \mu_2) \qquad 4.3.13(4),(5)
$$

$$= -(a^2 + \mu_2)(b + c) + (a + \mu_1)(b\beta + c\gamma),$$

$$
\begin{aligned}
b\beta + c\gamma &= (b + c)(\beta + \gamma) - b\gamma - c\beta \\
&= (b + c)(\beta + \gamma) + 2a^2 - a(\beta + \gamma) - b^2 - c^2 && 4.3.13(7),(8) \\
&= 2a^2 + (b + c - a)(\beta + \gamma) + 2\alpha^2 - (b + c)\alpha - 2\mu_1\alpha + 2\mu_2 && 4.3.13(3) \\
&= 2a^2 + 2\alpha^2 - (b + c)\alpha - 2\mu_1\alpha + 2\mu_2 + \epsilon(b + c) - \epsilon a && 4.3.6(1) \\
&\quad + (b + c)^2 - a^2 - (b + c)\alpha + a\alpha - (b + c)\tau + a\tau + (b + c)\mu_1 - a\mu_1.
\end{aligned}
$$

So we get

$$
\begin{aligned}
b\beta^2 &+ c\gamma^2 \\
&= a^3 + 2\alpha^2 a + a^2\alpha - 2(b + c)a\alpha - a^2(b + c) + (b + c)^2 a - \epsilon a^2 + \epsilon a(b + c) \\
&\quad + \tau(a(b + c) + a^2) + \mu_1(-a\alpha + (b + c)a - 2(b + c)\alpha + (b + c)^2 + 2\alpha^2 \\
&\quad + \epsilon(b + c) - \epsilon a + \tau(-(b + c) + a)) \\
&\quad + \mu_1^2(-2\alpha + (b + c) - a) - \mu_2((b + c) + 2a) + 2\mu_1\mu_2.
\end{aligned}
$$

Using the projection formula we get

$$
\begin{aligned}
\tilde{\pi}^*(\bar{\beta}) &= 3HP^2 + H^3 + H^2P - 2HH_2P - H^2H_2 + HH_2^2 - AH^2 + AHH_2 \\
&\quad + \mu_1(-HP + HH_2 - 2H_2P + H_2^2 + 2P^2 + AH_2 - AH) \\
&\quad + \mu_1^2(-2P + H_2 - H) + \mu_2(-H_2 + 2H) + 2\mu_1\mu_2.
\end{aligned}
$$

The formula for $\tilde{\pi}^*(\bar{\beta})$ is now obtained by applying the relations

$$
\begin{aligned}
H^3 &= \mu_1 H^2 - \mu_2 H + \mu_3, \\
AH^2 &= A(HP - P^2 + P\mu_1 - \mu_2). \qquad \square
\end{aligned}
$$

As we have determined $A^*(\widetilde{\mathrm{Hilb}}^3(\mathbf{P}(E)/X))$ in theorem 4.3.11, and its structure is in fact rather simple, this gives us a simple description of $A^*(\mathrm{Hilb}^3(\mathbf{P}(E)/X))$,
which is also very useful for computations. We now also want to describe this ring
by generators and relations. Because the relations are very complicated, we don't
want to state them all, but rather refer to [Göttsche (6)] for the list of all relations.

Theorem 4.4.7.

$$A^*(\mathrm{Hilb}^3(\mathbf{P}(E)/X)) = \frac{A^*(X)[\bar{H}, \bar{A}, \bar{h}, \bar{p}, \bar{a}, \bar{\alpha}, \bar{\beta}]}{(R_1, R_2, R_3, \ldots, R_{30})}$$

for suitable classes $R_1, R_2, R_3, \ldots, R_{30}$ *in* $A^*(X)[\bar{H}, \bar{A}, \bar{h}, \bar{p}, \bar{a}, \bar{\alpha}, \bar{\beta}]$, *which are all
listed in Satz 4.4.7 of [Göttsche (6)]. The relations in codimension at most three*

are

$$R_1 := -\bar{A}^2 - \bar{A}\bar{H} + 3\bar{a} - \bar{A}\mu_1,$$

$$R_2 := -\bar{H}\bar{p} + \bar{H}^3 + \bar{A}\bar{H}^2 - 4\bar{H}\bar{h} - \bar{H}\bar{a} + 3\bar{\alpha} - 3\bar{\beta}$$
$$\quad + \mu_1(5\bar{H}^2 + 4\bar{A}\bar{H} - 4\bar{h} - 6\bar{p} - 3\bar{a}) + \mu_1^2(10\bar{H} + 6\bar{A}) + \mu_2(-9\bar{H} + 3\bar{A})$$
$$\quad + 6\mu_1^3 - 18\mu_1\mu_2 + 9\mu_3,$$

$$R_3 := -\bar{A}\bar{h} + \bar{H}\bar{a} - 3\bar{\alpha} + 3\bar{a}\mu_1 - 3\bar{A}\mu_2,$$

$$R_4 := -\bar{A}\bar{p} + 3\bar{\alpha},$$

$$R_5 := -\bar{A}\bar{a} - \bar{H}\bar{a} + 3\bar{\alpha} - \bar{a}\mu_1.$$

Sketch of proof: The determination of the relations is a trivial but very extensive computation. We use theorem 4.4.6 and the relations $I_1, I_2, I_3, I_4, I_5, I_6$ of theorem 4.3.11, to express every element of the basis of $\widetilde{\pi}^*(A^*(\mathrm{Hilb}^3(\mathbf{P}(E)/X)))$ over $A^*(X)$ from proposition 4.4.4 as an $A^*(X)$-linear combination of elements of the basis of $A^*(\widetilde{\mathrm{Hilb}}^3(\mathbf{P}(E)/X))$ over $A^*(X)$ from the proof of proposition 4.3.10. For this we use the computer. Similarly we use proposition 4.4.6 and the relations $I_1, \ldots, I_6$ to express the images of

$$\bar{A}^2,$$

$$\bar{H}\bar{p}, \bar{A}\bar{h}, \bar{A}\bar{p}, \bar{A}\bar{a},$$

$$\bar{H}^4, \bar{A}\bar{H}^3, \bar{H}\bar{\beta}, \bar{A}\bar{\alpha}, \bar{A}\bar{\beta}, \bar{h}\bar{a}, \bar{p}^2, \bar{p}\bar{a}, \bar{a}^2,$$

$$\bar{H}^3\bar{h}, \bar{H}^3\bar{a}, \bar{H}^2\bar{\alpha}, \bar{h}\bar{\alpha}, \bar{h}\bar{\beta}, \bar{p}\bar{\beta}, \bar{a}\bar{\alpha}, \bar{a}\bar{\beta},$$

$$\bar{H}^3\bar{\alpha}, \bar{H}^2\bar{h}^2, \bar{h}^2\bar{p}, \bar{\alpha}^2, \bar{\alpha}\bar{\beta}, \bar{\beta}^2, \bar{H}^2\bar{h}\bar{p},$$

$$\bar{H}\bar{h}^3$$

under $\widetilde{\pi}^*$ as an $A^*(X)$-linear combination of the basis of $A^*(\widetilde{\mathrm{Hilb}}^3(\mathbf{P}(E)/X))$. Now we only have to solve a system of linear equations in order to get the relations. For this we use again the computer. As a result we get relations $R_1, \ldots, R_{30}$.

We still have to show that $R_1, \ldots, R_{30}$ generate all relations. For this we have to show that by using them we can express any monomial in $\bar{H}, \bar{A}, \bar{h}, \bar{p}, \bar{a}, \bar{\alpha}, \bar{\beta}$ as an $A^*(X)$-linear combination of the elements of the basis from proposition 4.4.4. To show this we use arguments similar to those in the end part of the proof of theorem 4.3.11. In the current case the arguments are however considerably more complicated and make use of the precise form of $R_1, \ldots, R_{30}$. We refer to the proof of Satz 4.4.7 in [Göttsche (6)] for the details. $\square$

In the rest of this section we look at an important special case of $\widetilde{\mathrm{Hilb}}^3(\mathbf{P}(E)/X))$. We put $G := Grass(d-2, d+1)$ and let $T := T_{3,d+1}$ be the tautological bundle over G.

Definition 4.4.8. Let $\text{Cop}^3(\mathbf{P}_d) \subset \text{Hilb}^3(\mathbf{P}_d) \times G$ be defined by

$$\text{Cop}^3(\mathbf{P}_d) := \left\{ (Z, E) \in \text{Hilb}^3(\mathbf{P}_d) \times G \mid Z \subset E \right\}.$$

Let $F \subset \mathbf{P}_d \times G$ be the incidence variety $F := \{ (x, E) \in \mathbf{P}_d \times G \mid x \in E \}$ with the projections

$$F$$

$$\swarrow {\scriptstyle p_1} \qquad \searrow {\scriptstyle p_2}$$

$$\mathbf{P}_d \qquad\qquad G.$$

In the same way as after definition 4.3.17 for $\widetilde{\text{Cop}}^3(\mathbf{P}_d)$ we see that there is a natural isomorphism

$$\psi : \text{Cop}^3(\mathbf{P}_d) \longrightarrow \text{Hilb}^3(\mathbf{P}(T)/G)$$

over G. The projection $\breve{p}_1 : \text{Cop}^3(\mathbf{P}_d) \longrightarrow \text{Hilb}^3(\mathbf{P}_d)$ is a birational morphism, as every subschmeme Z of length 3 of $\mathbf{P}_d$ is a subscheme of a plane. This plane is uniquely determined if Z does not lie on a line. In the same way as in the proof of remark 4.3.15 we see that the fibre $\breve{p}_1^{-1}(Z)$ over a point $Z \in Al^3(\mathbf{P}_d)$ is isomorphic to $\mathbf{P}_{d-2}$ and that the exceptional locus $\breve{p}_1^{-1}(Al^3(\mathbf{P}_d))$ is $Al^3(\mathbf{P}(T)/G) \cong \mathbf{P}(\text{Sym}^3(T_{2,T}))$. Here again $T_{2,T}$ is the tautological bundle of rank 2 over $\breve{\mathbf{P}}(T)$. This shows analogously to remark 4.3.15:

Remark 4.4.9. $\text{Cop}^3(\mathbf{P}_d)$ is the blow up of $\text{Hilb}^3(\mathbf{P}_d)$ along $Al^3(\mathbf{P}_d)$.

Definition 4.4.10. Let $\widehat{H}, \widehat{A}, \widehat{h}, \widehat{p}, \widehat{a}, \widehat{\alpha}, \widehat{\beta}, \widehat{\mu}_1, \widehat{\mu}_2, \widehat{\mu}_3 \in A^*(\text{Cop}^3(\mathbf{P}_d))$ be the classes

$$\widehat{A} := \left[\left\{ (Z, E) \in \text{Cop}^3(\mathbf{P}_d) \mid Z \text{ lies on a line } \right\} \right],$$

$$\widehat{H} := \left[\left\{ (Z, E) \in \text{Cop}^3(\mathbf{P}_d) \mid Z \text{ intersects a fixed hyperplane} \right\} \right],$$

$$\widehat{h} := \left[\left\{ (Z, E) \in \text{Cop}^3(\mathbf{P}_d) \,\middle|\, \begin{array}{c} Z \text{ intersects a fixed 2-codimensional} \\ \text{linear subspace} \end{array} \right\} \right],$$

$$\widehat{p} := \left[\left\{ (Z, E) \in \text{Cop}^3(\mathbf{P}_d) \,\middle|\, \begin{array}{c} \text{the line through one of the subschemes } Z' \subset Z \\ \text{of length 2 intersects two different} \\ \text{fixed 2-codimensional linear subspaces} \end{array} \right\} \right],$$

$$\widehat{a} := \left[\left\{ (Z, E) \in \text{Cop}^3(\mathbf{P}_d) \,\middle|\, \begin{array}{c} Z \text{ lies on a line intersecting a fixed} \\ \text{2-codimensional linear subspace} \end{array} \right\} \right],$$

$$\widehat{\alpha} := \left[\left\{ (Z, E) \in \mathrm{Cop}^3(\mathbf{P}_d) \,\middle|\, \begin{array}{c} Z \text{ lies on a line} \\ \text{intersecting two different 2-codimensional} \\ \text{linear subspaces} \end{array} \right\} \right],$$

$$\widehat{\beta} := \left[\left\{ (Z, E) \in \mathrm{Cop}^3(\mathbf{P}_d) \,\middle|\, \begin{array}{c} \text{the line through one of the subschemes } Z' \subset Z \\ \text{of length 2 intersects two different} \\ \text{fixed 2-codimensional linear subspaces,} \\ res(Z', Z) \text{ lies on a fixed hyperplane} \end{array} \right\} \right],$$

$$\widehat{\mu}_1 := \left[\left\{ (Z, E) \in \mathrm{Cop}^3(\mathbf{P}_d) \,\middle|\, \begin{array}{c} E \text{ intersects a fixed} \\ \text{3-codimensional linear subspace} \end{array} \right\} \right],$$

$$\widehat{\mu}_2 := \left[\left\{ (Z, E) \in \mathrm{Cop}^3(\mathbf{P}_d) \,\middle|\, \begin{array}{c} E \text{ has a one-dimensional intersection with a fixed} \\ \text{2-codimensional linear subspace} \end{array} \right\} \right],$$

$$\widehat{\mu}_3 := \left[\left\{ (Z, E) \in \mathrm{Cop}^3(\mathbf{P}_d) \,\middle|\, E \text{ lies on a fixed hyperplane} \right\} \right].$$

From the definitions we get:

Remark 4.4.11.

$$\psi^*(\bar{A}) = \widehat{A}, \ \psi^*(\bar{H}) = \widehat{H},$$
$$\psi^*(\bar{h}) = \widehat{h}, \ \psi^*(\bar{p}) = \widehat{p}, \ \psi^*(\bar{a}) = \widehat{a},$$
$$\psi^*(\bar{\alpha}) = \widehat{\alpha}, \ \psi^*(\bar{\beta}) = \widehat{\beta},$$
$$\psi^*(\mu_1) = \widehat{\mu}_1, \ \psi^*(\mu_2) = \widehat{\mu}_2, \ \psi^*(\mu_3) = \widehat{\mu}_3.$$

So theorem 4.4.6 describes the Chow ring of $\mathrm{Cop}^3(\mathbf{P}_d)$ in terms of classes describing the position of subschemes relative to linear subspaces of $\mathbf{P}_d$.

In the case of $\mathrm{Cop}^3(\mathbf{P}_3)$ we get in particular:

$$\mu := \hat{\mu}_1 = \left[\left\{ (Z, E) \in \mathrm{Cop}^3(\mathbf{P}_d) \,\middle|\, E \text{ contains a fixed point} \right\} \right],$$
$$\hat{\mu}_2 = \mu^2,$$
$$\hat{\mu}_3 = \mu^3.$$

We can now use theorem 4.4.6 to compute the intersection tables with the help of the computer. We keep in mind that for $u \in A^i(\mathrm{Cop}^3(\mathbf{P}_3))$, $v \in A^{9-i}(\mathrm{Cop}^3(\mathbf{P}_3))$ the intersection numbers $u \cdot v$ and $\widetilde{\pi}^*(u) \cdot \widetilde{\pi}^*(v)$ are related by $\widetilde{\pi}^*(u) \cdot \widetilde{\pi}^*(v) = 3u \cdot v$ and obtain the following tables:

$A^1 \times A^8$:

	$\hat{H}$	$\hat{A}$	μ
$\hat{h}^3\mu^2$	3	2	1
$\hat{H}\hat{h}^2\mu^3$	1	1	
$\hat{H}\hat{h}\hat{p}\mu^3$	1		

$A^2 \times A^7$:

	$\hat{H}^2$	$\hat{H}\hat{A}$	$\hat{h}$	$\hat{a}$	$\hat{p}$	$\hat{H}\mu$	$\hat{A}\mu$	μ^2
$\hat{h}^3\mu$	6	3		2	6	3	2	1
$\hat{H}\hat{h}^2\mu^2$	7	6	3	2	4	1	1	
$\hat{H}\hat{h}\hat{p}\mu^2$	15	6	4		4	1		
$\hat{H}^2\hat{h}\mu^3$	3	3	1	1	1			
$\hat{H}^2\hat{a}\mu^3$	6	-3	1	-1				
$\hat{h}^2\mu^3$	1	1	1					
$\hat{H}\hat{a}\mu^3$	1	-1						
$\hat{h}\hat{p}\mu^3$	1							

$A^3 \times A^6$:

	$\hat{H}^3$	$\hat{h}\hat{H}$	$\hat{H}^2\hat{A}$	$\hat{H}\hat{a}$	$\hat{\alpha}$	$\hat{\beta}$	$\hat{H}^2\mu$	$\hat{H}\hat{A}\mu$	$\hat{h}\mu$	$\hat{a}\mu$	$\hat{p}\mu$	$\hat{H}\mu^2$	$\hat{A}\mu^2$	μ^3
$\hat{h}^3$	6					6	6	3		2	6	3	2	1
$\hat{H}\hat{h}^2\mu$	20	6	13	6	2	6	7	6	3	2	4	1	1	
$\hat{H}\hat{h}\hat{p}\mu$	66	18	27	6		13	15	6	4		4	1		
$\hat{H}^2\hat{h}\mu^2$	25	7	22	9	2	5	3	3	1	1	1			
$\hat{H}^2\hat{a}\mu^2$	40	9	-10	-7	-2	2	6	-3	1	-1				
$\hat{h}^2\mu^2$	7	3	6	2			1	1	1					
$\hat{H}\hat{a}\mu^2$	12	2	-7	-2			1	-1						
$\hat{h}\hat{p}\mu^2$	15	4	6			2	1							
$\hat{H}^3\mu^3$	15	3	15	6	1	3								
$\hat{h}\hat{H}\mu^3$	3	1	3	1										
$\hat{H}^2\hat{A}\mu^3$	15	3	3	-3	-1	1								
$\hat{H}\hat{a}\mu^3$	6	1	-3	-1										
$\hat{\alpha}\mu^3$	1		-1											
$\hat{\beta}\mu^3$	3		1											

$A^4 \times A^5$:

	$\hat{H}^2\hat{h}$	$\hat{H}^2\hat{a}$	$\hat{h}^2$	$\hat{H}\hat{a}$	$\hat{h}\hat{p}$	$\hat{H}^3\mu$	$\hat{h}\hat{H}\mu$	$\hat{H}^2\hat{A}\mu$	$\hat{H}\hat{a}\mu$	$\hat{a}\mu$	$\hat{\beta}\mu$	$\hat{H}^2\mu^2$	$\hat{H}\hat{A}\mu^2$	$\hat{h}\mu^2$	$\hat{a}\mu^2$	$\hat{p}\mu^2$	$\hat{H}\mu^3$	$\hat{A}\mu^3$
$\hat{H}\hat{h}^2$	6				6	20	6	13	6	2	6	7	6	3	2	4	1	1
$\hat{H}\hat{h}\hat{p}$	26		6		24	66	18	27	6		13	15	6	4			4	1
$\hat{H}^2\hat{h}\mu$	20	22	6	9	18	25	7	22	9	2	5	3	3	1	1	1		
$\hat{H}^2\hat{a}\mu$	22	−16	6	−8	6	40	9	−10	−7	−2	2	6	−3	1	−1			
$\hat{h}^2\mu$	6	6		2	6	7	3	6	2			1	1	1				
$\hat{H}\hat{a}\mu$	9	−8	2	−2		12	2	−7	−2			1	−1					
$\hat{h}\hat{p}\mu$	18	6	6		8	15	4	6				2	1					
$\hat{H}^3\mu^2$	25	40	7	12	15	15	3	15	6	1	3							
$\hat{h}\hat{H}\mu^2$	7	9	3	2	4	3	1	3	1									
$\hat{H}^2\hat{A}\mu^2$	22	−10	6	−7	6	15	3	3	−3	−1	1							
$\hat{H}\hat{a}\mu^2$	9	−7	2	−2		6	1	−3	−1									
$\hat{a}\mu^2$	2	−2				1		−1										
$\hat{\beta}\mu^2$	5	2			2	3		1										
$\hat{H}^2\mu^3$	3	6	1	1	1													
$\hat{H}\hat{A}\mu^3$	3	−3	1	−1														
$\hat{h}\mu^3$	1	1	1															
$\hat{a}\mu^3$	1	−1																
$\hat{p}\mu^3$	1																	

Bibliography

Altman, A. Kleiman, S.
(1) Compactifying the Picard scheme, Adv. in Math. **35** (1980), 50-112.

Andrews, G. E.
(1) The Theory of Partitions, Encyclopedia of Mathematics and its Applications, Addison-Wesley, Reading, Massachusetts 1976.

Arrondo, E., Sols, I., Speiser, R.
(1) Global Moduli for Contacts, Preprint Oktober 1992.

Avritzer, D., Vainsencher, I.
(1) $Hilb^4 P_2$, Enumerative Geometry, Proc Sitjes 1987, S. Xambó-Descamps, ed., Lecture Notes in Math. **1436**, Springer-Verlag, Berlin Heidelberg 1990, 30-59.

Barth, W., Peters, C. Van de Ven, A.
(1) Compact complex surfaces, Ergebnisse der Mathematik und ihrer Grenzgebiete 3. Folge, Band 4, Springer-Verlag, Berlin Heidelberg New York Tokyo 1984.

Beauville, A.
(1) Variétés kähleriennes dont la première classe de Chern est nulle, J. Diff. Geometry **18** (1983), 755-782.
(2) Some remarks on Kähler manifolds with $c_1 = 0$, Classification of algebraic and analytic manifolds Katata 1982, Progr. Math. **39**, Birkhäuser Boston, Boston, Mass. 1983, 1-26.
(3) Variétés kähleriennes avec $c_1 = 0$, Geometry of $K3$ surfaces: moduli and periods, Palaiseau 1981/1982, Astérisque **126** (1985), 181-192.

Beltrametti, M., Sommese A. J.
(1) Beltrametti, M., Sommese, A. J., Zero cycles and k-th order embeddings of smooth projective surfaces, Cortona proceedings Problems in the Theory of Surfaces and their Classification, Symposia Mathematica XXXII, INDAM, Academic Press, London San Diego New York Boston Sydney Tokyo Toronto 1991, 33-44

Beltrametti, M., Francia, P., Sommese, A. J.
(1) On Reider's method and higher order embeddings, Duke Math. Journal **58** (1989), 425-439.

Bialynicki-Birula, A.
(1) Some theorems on actions of algebraic groups, Annals of Math. **98** (1973), 480-497.
(2) Some properties of the decompositions of algebraic varieties determined by actions of a torus, Bull. de l'Acad. Polonaise des Sci., Série des sci. math. astr. et phys. **24** (1976), 667-674.

Bialynicki-Birula, A., Sommese, A. J.
(1) Quotients by $\mathbf{C}^*$ and $SL(2, \mathbf{C})$ actions, Transactions of the American Math. Soc. **279** (1983), 45-89.

Briançon, J.

(1) Description de $Hilb^n C\{x,y\}$, Invent. Math. **41** (1977), 45-89.

Catanese, F., Göttsche, L.

(1) d-very-ample line bundles and embeddings of Hilbert schemes of 0-cycles, Manuscripta Math. **68** (1990),337-341.

Cheah, J.

(1) The Hodge numbers of the Hilbert scheme of points on a smooth projective surface, preprint 1993.

Chow, W.-L., Van der Waerden, B. L.

(1) Über zugeordnete Formen und algebraische Systeme von algebraischen Mannigfaltigkeiten, Math. Annalen **113** (1937), 692-704.

Colley, S. J., Kennedy, G.

(1) A higher order contact formula for plane curves, Comm. Algebra **19** (1991) no. 2, 479-508.

(2) Triple and quadruple contact of plane curves, Enumerative algebraic geometry (Copenhagen 1989), Contemp. Math. **123**, Amer. Math. Soc., Providence, RI, 1991, 31-59.

Collino, A.

(1) Evidence for a conjecture of Ellingsrud and Strømme on the Chow Ring of $Hilb^d(\mathbf{P}^2)$, Illinois J. of Math. **32** (1988), 171-210.

Collino, A., Fulton, W.

(1) Intersection Rings of Spaces of Triangles, Société Mathématique de France, Memoire **38** (1989), 75-117.

Deligne, P.

(1) La conjecture de Weil, I, Inst. Hautes Etudes Sci. Publ. Math. **43** (1974), 273-307.

Digne, F.

(1) Shintani descent and $\mathcal{L}$ functions on Deligne-Lustig varieties, Proceedings of Symposia in pure Mathematics of the AMS, Volume **47** part 1, The Arcata Conference on Representations of Finite Groups 1986, 297-320.

Dixon, L., Harvey, J. Vafa, C., Witten, E.

(1) Strings on orbifolds I, Nucl. Phys. B **261** (1985), 678-686.

(2) Strings on orbifolds II, Nucl. Phys. B **274** (1986), 285-314.

Douady, A.

(1) Le problème des modules pour les sous-espaces analytiques d'un espace analytique donné, Ann. Inst. Fourrier (Grenoble) **16**-1 (1966), 1-95.

Eisenbud, D., Harris, J.

(1) Schemes: The language of moderne algebraic geometry, Wadsworth & Brooks/Cole Advanced Books & Software, Pacific Grove, California 1992.

Elencwajg, G., Le Barz, P.
(1) Une Base de $Pic(Hilb^3\mathbf{P}^2)$, C. R. Acad. Sci. Paris **297** (1983), 175-178.
(2) Détermination de l'anneau de Chow de $Hilb^3\mathbf{P}^2$, C. R. Acad. Sci. Paris **301** (1985), 635-638.
(3) L'anneau de Chow des triangles du plan, Comp. Math **71** (1989), 85-119.
(4) Applications énumératives du calcul de $CH(Hilb^3\mathbf{P}^2)$, preprint Univ. Nice 1985.
(5) Explicit Computations in $Hilb^3\mathbf{P}^2$, Proc. Algebraic Geometry Sundance 1986, Holme, A. and Speiser, R., eds, Lecture Notes in Math. **1311**, Springer-Verlag, Berlin Heidelberg 1988, 76-100.

Ellingsrud, G.
(1) Another proof of the irreducibility of the punctual Hilbert scheme of a smooth surface, preprint 1992.

Ellingsrud, G., Strømme, S. A.
(1) On the homology of the Hilbert scheme of points in the plane, Invent. Math. **87** (1987), 343-352.
(2) On a cell decomposition of the Hilbert scheme of points in the plane, Invent. Math **91** (1988), 365-370.
(3) On generators for the Chow ring of fine moduli spaces on $\mathbf{P}^2$, preprint 1989.
(4) Towards the Chow ring of the moduli space for stable sheaves on $\mathbf{P}^2$ with $c_1 = 1$, preprint 1991.
(5) Towards the Chow ring of the Hilbert scheme of $\mathbf{P}_2$, J. reine angew. Math. **441** (1993), 33-44.

Fantechi, B., Göttsche, L.
(1) The cohomology ring of the Hilbert scheme of 3 points on a smooth projective variety, J. reine angew. Math. **439** (1993), 147-158.

Fogarty, J.
(1) Algebraic families on an algebraic surface, American Journal of Math. **90** (1968), 511-521.
(2) Algebraic families on an algebraic surface II, the Picard scheme of the punctual Hilbert scheme, Amercan Journal of Math. **95** (1973), 660-687.

Fulton, W.
(1) Intersection Theory, Ergebnisse der Mathematik und ihrer Grenzgebiete, Springer-Verlag, Berlin Heidelberg New York Tokyo 1984.
(2) Introduction to intersection theory in algebraic geometry, CBMS Regional Conf. Ser. in Math. vol. 44, AMS, Providence, R. I., 1984.

Göttsche, L.
(1) Die Betti-Zahlen des Hilbert-Schemas für Unterschemata der Länge n auf einer glatten Fläche, Diplomarbeit, Bonn, Juli 1988.
(2) The Betti numbers of the Hilbert scheme of points on a smooth projective surface, Math. Ann. **286** (1990), 193-207.
(3) Identification of very ample line bundles on $S^{[r]}$, Appendix to: Beltrametti,

M., Sommese, A. J., Zero cycles and k-th order embeddings of smooth projective surfaces, Cortona proceedings Problems in the Theory of Surfaces and their Classification, Symposia Mathematica XXXII, INDAM, Academic Press, London San Diego New York Boston Sydney Tokyo Toronto 1991, 44-48.

(4) Betti numbers for the Hilbert function strata of the punctual Hilbert scheme in two variables, Manuscripta Math. **66** (1990), 253-259.

(5) The Betti numbers of higher order Kummer varieties of surfaces, preprint 1989.

(6) Hilbertschemata nulldimensionaler Unterschemata glatter Varietäten, thesis 1991.

Göttsche, L., Soergel W.

(1) Perverse sheaves and the cohomology of Hilbert schemes of smooth algebraic surfaces, Math. Ann. **296** (1993), 235-245.

Granger, M.

(1) Géométrie des schémas de Hilbert ponctuels, Mém. Soc. Math. France **8** (1983).

Griffiths, P., Harris, J.

(1) Principles of Algebraic Geometry, Wiley and Sons, New York (1978).

Grothendieck, A.

(1) Techniques de construction et théorèmes d'existence en géométrie algébrique IV: Les schémas de Hilbert, Séminaire Bourbaki exposé **221** (1961), IHP, Paris.

Harder, G. , Narasimhan, M. S.

(1) On the Cohomology Groups of Moduli Spaces of Vector Bundles on Curves, Math. Ann. **212** (1975), 215-248.

Hartshorne, R.

(1) Connectedness of the Hilbert scheme, Publ. Math. IHES **29** (1966), 261-304.

(2) Algebraic Geometry, Graduate Texts in Mathematics 52, Springer-Verlag, New York Heidelberg Berlin (1977).

Hirschowitz, A.

(1) Le group de Chow équivariant, C. R. Acad. Sci. Paris **298** (1984).

Hirzebruch, F.

(1) Topological Methods in Algebraic Geometry, Grundlehren der mathematischen Wissenschaften 131, third Edition, Springer-Verlag, Berlin Heidelberg New York 1978.

Hirzebruch, F. with Berger, T. Jung, R.

(1) Manifolds and Modular Forms, Vieweg, Braunschweig/Wiesbaden, 1992.

Hirzebruch, F., Höfer, T.

(1) On the Euler number of an orbifold, Math. Ann **286** (1990), 255-260.

Iarrobino, A.

(1) Reducibility of the families of 0-dimensional schemes on a variety, Invent. Math. **15** (1972), 72-77.

(2) Punctual Hilbert schemes, Bull. Amer. Math. Soc. **78** (1972), 819-823.
(3) An algebraic fibre bundle over $\mathbf{P}_1$, that is not a vector bundle, Topology **12** (1973), 229-232.
(4) Punctual Hilbert schemes, Mem. Amer. Math. Soc. **188** (1977).
(5) Hilbert scheme of points: an Overview of last ten Years, Proceedings of Symposia in pure Mathematics of the AMS, Volume **46** part 2, Algebraic Geometry 1985, 297-320.

Iarrobino, A., Yameogo, J.
(1) Cohomology groups of the family G_T of graded algebra quotients of $k[[x, y]]$ having Hilbert function T; and the hook differences of partitions with diagonal lengths T, preprint 1990.
(2) Partitions of diagonal lengths T and ideals in $k[[x, y]]$, preprint 1991.

Keel, S.
(1) Functorial construction of Le Barz's triangle space with applications, Trans. Amer. Math. Soc. **335** (1993), 213-229.

Kirwan, F. C.
(1) Cohomology of Quotients in Symplectic and Algebraic Geometry, Princeton University Press, Princeton, New Jersey 1984.

Kleiman, S. L.
(1) The transversality of the general translate, Compositio Math. **28** (1974), 287-297.
(2) Multiple-point formulas I: iteration, Acta Math. **147** (1981), 287-297.
(3) Multiple point formulas II: the Hilbert scheme, Enumerative Geometry, Proc Sitjes 1987, S. Xambó-Descamps, ed., Lecture Notes in Math. **1436**, Springer-Verlag, Berlin Heidelberg 1990, 101-138.
(4) Rigorous foundation of Schubert's enumerative calculus, Math. Development from Hilbert problems, Proceedings of Symposia in pure Mathematics of the AMS, Volume **28** (1983), 445-482.
(5) Intersection Theory and enumerative Geometry, Proceedings of Symposia in pure Mathematics of the AMS, Volume **46** part 2, Algebraic Geometry 1985, 321-370.

Le Barz, P.
(1) Géométrie énumerative pour les multisécantes, Variétés Analytiques Compacts, Proc. Conf. Nice 1977, Lecture Notes in Math. **683**, Springer-Verlag, Berlin Heidelberg 1978, 116-167.
(2) Validité de certaines formules de géométrie énumérative, C. R. Acad. Sci. Paris **289** (1979), 755-759.
(3) Quadrisécantes d'une surface de $\mathbf{P}^5$, C. R. Acad. Sci. Paris **291** (1980), 639-642.
(4) Formules pour les multisécantes des surfaces, C. R. Acad. Sci. Paris **292** (1981), 797-800.
(5) Formules multisécantes pous les courbes gauches quelconques, Enumerative and

classical algebraic Geometry, Nice 1981, Prog. in Math. 24, Birkhäuser 1982, 165-197.

(6) Platitude et non-platitude de certaines sous-schémas de $Hilb^k\mathbf{P}^N$, Journal für die Reine und Angewandte Mathematik **348** (1984), 116-134.

(7) Contribution des droites d'une surface à ses multisécantes, Bull. Soc. Math. France **112** (1984), 303-324.

(8) Quelques calculs dans les variétés d'alignements, Adv. in Math. **64** (1987), 87-117.

(9) Formules pour les trisécantes des surfaces algébriques. Enseign. Math. **33** no 1-2 (1987), 1-66.

(10) La variété des triplets complets, Duke Math. Journal **57** (1988), 925-946.

(11) Quelques formules multisécantes pour les surfaces, Enumerative Geometry, Proc Sitjes 1987, S. Xambó-Descamps, ed., Lecture Notes in Math. **1436**, Springer-Verlag, Berlin Heidelberg 1990, 151-188.

Macdonald, I. G.

(1) The Poincaré polynomial of a symmetric product, Proc. Camb. Phil. Soc. **58** (1962), 563-568.

Mallavibarrena, R.,

(1) Les groupes de Chow de $Hilb^4\mathbf{P}^2$, C. R. Acad. Sci. Paris **303**, I13 (1986).

(2) Validité de la formule classique des trisécantes stationaires, C. R. Acad. Sci. Paris **303**, I16 (1986).

(3) El Método de las bases de los grupos de Chow de $Hilb^d\mathbf{P}^2$ en geometria enumerativa, Thesis 1987.

Mallavibarrena, R., Sols, I.

(1) Bases for the homology groups of the Hilbert scheme of points in the plane, Comp. Math. **74** (1990), 169-202.

Matsumura, H.

(1) Commutative Algebra, W. A. Benjamin, Inc., New York 1970.

Mazur, B.

(1) Eigenvalues of Frobenius acting on algebraic varieties over finite fields, Proceedings of Symposia in Pure Mathematics Vol. **29**, Algebraic Geometry, Arcata 1974, 231-261.

Milne, J. S.

(1) Etale Cohomology, Princeton Math. Series **33**, Princeton University Press, Princeton 1980.

Mumford D.

(1) Lectures on curves on an algebraic surface, Annals of Math. Studies vol. **59**, Princeton 1966.

(2) The Red Book on Varieties and Schemes, Lecture Notes in Mathematics **1358**, Springer-Verlag, Berlin Heidelberg 1988.

Mumford, D. Fogarty, J.
(1) Geometric invariant theory, Ergebnisse der Mathematik und ihrer Grenzgebie-
    te, second edition, Springer-Verlag, New York Heidelberg Berlin 1982.

Reider, I.
(1) Vector bundles of rank 2 and linear systems on algebraic surfaces, Ann. of
    Math. **127** (1988), 309-316.

Roan, S. S.
(1) On the generalization of Kummer surfaces, J. Diff. Geom. **30** (1989) 523-537.

Roberts, J.
(1) Old and new results about the triangle variety, Proc. Algebraic Geometry
    Sundance 1986, Holme, A. and Speiser, R., eds, Lecture Notes in Math. **1311**,
    Springer-Verlag, Berlin Heidelberg 1988, 197-219.

Roberts, J. Speiser, R.
(1) Schubert's enumerative geometry of triangles from a modern viewpoint, Alge-
    braic Geometry, Proc. Conf. Chicago Circle 1980, Lecture Notes in Math. **862**,
    Springer Verlag, Berlin Heidelberg 1981, 272-281.

(2) Enumerative geometry of triangles I, Comm in Alg. **12**(9-10) (1984), 1213-1255.
(3) Enumerative geometry of triangles II, Comm in Alg. **14**(1) (1986), 155-191.
(4) Enumerative geometry of triangles III, Comm in Alg. **15**(9) (1987), 1929-1966.

Rosselló Llompart, F.
(1) Les groupes de Chow de quelques schémas qui paramétrisent des points
    coplanaires, C. R. Acad. Sci. Paris Sér. I Math. **303** (1986), 363-366.
(2) The Chow-Ring of $Hilb^3 \mathbf{P}^3$, Enumerative Geometry, Proc Sitjes 1987, S.
    Xambó Descamps, ed., Lecture Notes in Math. **1436**, Springer-Verlag, Berlin
    Heidelberg 1990, 225-255.

Rosselló Llompart, F., Xambó Descambs, S.
(1) Computing Chow groups, Proc. Algebraic Geometry Sundance 1986, Holme,
    A. and Speiser, R., eds, Lecture Notes in Math. **1311**, Springer-Verlag, Berlin
    Heidelberg 1988, 220-234.
(2) Chow groups and Borel-Moore schemes, Ann. Math. Pura Appl. (4) **160**
    (1991), 19-40.

Schubert, H.
(1) Kalkül der abzählenden Geometrie, Teubner, Leibzig 1879, reprinted by
    Springer-Verlag, Berlin 1979.
(2) Anzahlgeometrische Behandlung des Dreiecks, Math. Ann. **17** (1880), 153-212.

Speiser, R.
(1) Enumerating contacts, Proceedings of Symposia in pure Mathematics of the
    AMS, Volume **46** part 2, Algebraic Geometry 1985, 401-418.

Semple, J. G.
(1) Some investigations in the geometry of curve and surface elements, Proc. London Math. Soc. 4(3) (1954), 24-49.
(2) The triangle as a geometric variable, Mathematica 1 (1954), 80-88.

Steenbrink, J. H. M.
(1) Mixed Hodge Structures on the vanishing cohomology, Nordic Summer School, Symposium in Mathematics, Oslo 1970, Sijthoff and Noordhoff, Alphen an den Rijn 1977, (525-563).

Tyrrell, J. A.
(1) On the enumerative geometry of triangles, Mathematica 6 (1959), 158-164.

Tyurin, A. N.
(1) Cycles, curves and vector bundles on an algebraic surface, Duke Math. J. 54 (1987), 1-26.

Zagier, D.
(1) Equivariant Pontrjagin classes and applications to orbit spaces, Lecture Notes in Math. 290, Springer-Verlag, Berlin Heidelberg New York 1972.
(2) Note on the Landweber-Stong elliptic genus, Elliptic Curves and Modular Forms in Algebraic Topology, Proceedings Princeton 1986, P.S. Landweber (Ed.), Lecture Notes in Mathematics 1326, Springer-Verlag, Berlin Heidelberg New York 1988, 216-224.

Index

Index of notations

Printing: Weihert-Druck GmbH, Darmstadt
Binding: Buchbinderei Schäffer, Grünstadt